Property of
Research & Development Library
Cingular Wireless

Andreas Springer · Robert Weigel

UMTS

Springer

Berlin
Heidelberg
New York
Barcelona
Hong Kong
London
Milan
Paris
Tokyo

Engineering 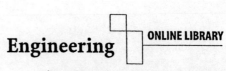 ONLINE LIBRARY

http://www.springer.de/engine/

Andreas Springer · Robert Weigel

UMTS

The Physical Layer of the Universal Mobile Telecommunications System

With 139 Figures

 Springer

Dr. Andreas Springer
Professor Robert Weigel
University of Linz
Institute for Communications
and Information Engineering
Altenbergerstr. 69
4040 Linz
Austria
e-mails: springer@mechatronik.uni-linz.ac.at
r. weigel@ieee.org

ISBN 3-540-42162-9 Springer-Verlag Berlin Heidelberg NewYork

Die Deutsche Bibliothek – CIP-Einheitsaufnahme

Springer, Andreas:
UMTS: The Physical Layer of the Universal Mobile Telecommunications System / Andreas Springer,
Robert Weigel. – Berlin; Heidelberg; New York; Barcelona, Hong Kong; London; Milan; Paris; Tokyo:
Springer, 2002
ISBN 3-540-42162-9

Springer-Verlag is a company in the BertelsmannSpringer publishing group
http://www.springer.de
© Springer-Verlag Berlin Heidelberg New York 2002
Printed in Germany

Typesetting: Camera-ready copy from authors
Cover-Design: de'blik, Berlin
Printed on acid-free paper   SPIN: 10788804   62/3020/kk   5 4 3 2 1 0

Preface

It took more than 15 years of research, standardization work, and development with an enormous effort of manpower to bring third generation (3G) wireless communication systems to life. The first research work on 3G systems started around 1988 [1]. At this time the striking success of second generation (2G) systems, especially of GSM (Global System for Mobile Communications), was not yet evident. A substantial part of these early research activities took place in Europe and was sponsored by the European Commission in the course of research programs such as: Research and Development of Advanced Communications Technologies in Europe (RACE-1, RACE-2) and Advanced Communications Technology and Services (ACTS) [2]. Even before these activities, 3G systems were considered in 1992 at the WARC (World Administrative Radio Conference), where 230 MHz of spectrum around 2 GHz was identified for 3G, and in standardization bodies like the ITU (International Telecommunications Union) from a global perspective and ETSI (European Telecommunications Standards Institute) in Europe. At the present time 3G networks are deployed or are already operating (e.g., in Japan the first commercial 3G system started its service in October 2001). Unfortunately, the initial idea to create one single 3G standard to allow for seamless world-wide roaming could not be realized. However, three of the five members of the so-called IMT-2000 (International Mobile Telecommunications, the official acronym for 3G systems) family of standards are based on Wideband-CDMA (Code Division Multiple Access). The European contributions to IMT-2000 are named UMTS (Universal Mobile Telecommunications System) with its two modes FDD (Frequency Division Duplex) and TDD (Time Division Duplex). The first operational 3G system is based on UMTS FDD, as will be the first systems in Europe, scheduled to start around the end of 2002.

While a vast amount of research and development has been devoted to networking, protocol, and signal processing issues for 3G, the analog and RF (radio frequency) building blocks and their necessary performance have received considerably less attention in the course of IMT-2000. This book tries make up for a small fraction of this gap. The idea for this book originated after a short course on the UMTS standard was developed at the Institute for Communications and Information Engineering at the Johannes Kepler

University Linz, Austria, and held for the first time in February 2000 at the 2000 International Zurich Seminar on Broadband Communications in Zurich, Switzerland. During our work in the course of the LEMON project, funded by the European Commission under contract number IST-1999-11081, which is devoted to the implementation of a 3G transceiver chip in CMOS (complementary metal-oxide semiconductor) technology, and in numerous discussions with engineers and researchers, it turned out that there was a considerable need from system- and hardware engineers to learn about UMTS, its basic principles, the underlying CDMA technology, and the physical layer specifications. We adopted a similar order for the organization of this book.

After a very general introduction into 3G communications in chapter 1, chapter 2 deals with the fundamentals of mobile communications, including basic concepts, digital modulation, the mobile radio channel and some RF basics. Similarly to chapters 3 and 4, all these topics are covered at a very basic level just to give the reader enough information to better understand the following description of the physical layer of UMTS and the related front-end requirements in the second part of this book. In chapter 3, spread spectrum techniques with an emphasis on direct sequence spread spectrum is covered, which is the basis for the code division multiple access technique (on which UMTS is based) described in chapter 4. The second part of the book is devoted to a detailed description of the physical layer of UMTS. Chapter 5 provides an introduction to the concept behind 3G systems, takes a short look at market aspects and spectrum issues, and the standardization of IMT-2000 is covered. Chapter 6 deals with the UMTS FDD mode in detail. After a short explanation of the architecture of the whole system, details on the transport and physical channels, spreading and modulation, physical layer procedures, the compressed mode, and the required RF characteristics are described. In chapter 7, the same is repeated for the UMTS TDD mode. Finally, chapter 8 reviews receiver and transmitter architectures with respect to their suitability for UMTS and reviews some of the published work on RF front-ends for UMTS. In the last section of chapter 8, possible trends for the design and development of future equipment for 3G systems are identified.

Chapters 6 and 7 closely follow the relevant standards documents but attempt to compress and compile the vast amount of information, sometimes spread over several documents, in a way most useful and convenient for the reader. All information is – if already available – based on release 4 of the UMTS specifications. This release is not yet finalized and current equipment implementations are based on release 99. However, we felt that it is worth basing this book on release 4, even though it is still under development, for two reasons. First, most future products must be built according to release 4, and thus the information provided in chapters 6 and 7 will not become outdated as fast as it would if we had based our text on release 99. Second, there are only minor differences between the two releases for most of the presented material. The only major difference is the introduction of the 1.28 Mcps option for the TDD mode.

The authors would like to acknowledge many people who have contributed to this book. First of all the work of Dr. Mario Huemer and Dipl.-Ing. Dieter Pimingsdorfer, both of DICE/Infineon, Linz, and Dr. Linus Maurer of our Institute, who were our co-authors on the short course on UMTS out of which this book has been developed, is highly appreciated. Especially appreciated are the numerous discussions with Dr. Maurer that shed light on to many details of UMTS. Our colleagues Dipl.-Ing. Bernd Adler and Zdravko Boos, MEE, both of Infineon, Munich, always tried to direct our focus towareds practical implementation problems. Dr. Harald Pretl and Dr. Werner Schelmbauer, both of DICE/Infineon, Linz, gave us helpful support with all kinds of circuit-related issues. Ralf Rudersdorfer helped with the preparation of the graphics. Ms. Jantzen and Dr. Merkle of Springer-Verlag are acknowledged for their support and co-operation. Finally, the authors would like to thank their colleagues at the Institute for Communications and Information Engineering at the Johannes Kepler University Linz for their support.

Linz, March 2002   *A. Springer and R. Weigel*

Contents

1. Introduction

Mobile radio communication systems have been successfully deployed throughout the world since the early 1980s. The first generation analog systems supported mainly voice telephony and paved the way into a wireless world. With second generation (2G) digital systems the cellular phone became a widespread tool used in our everyday life. Today, the market penetration of cellular phones already exceeds 80% in some countries and makes mobile telephony an outstanding telecommunications success. The mobile communications mass market has fueled the growth of a huge wireless industry comprising IC and equipment manufacturers, network operators, software companies, service and content providers, etc. This big economic success took place despite the very limited wireless communications possibilities (low data rates and mostly circuit-switched connections) compared to traditional wire line services. At the beginning of the wireless era user mobility was the new and fascinating feature which attracted a large number of subscribers. The trend towards mobility for the user is nowadays accompanied by the demand for access to digital data as a second trend. Until now, these two needs have been satisfied by two different solutions, namely mobile communication networks and the Internet. Driven by the Internet, the amount of electronically accessible information is growing exponentially. Digitized information from paper, film and tape archives and new multimedia information content based on image, video, voice, and text all exist today and such information is being delivered on a large scale over the fixed Internet. The convergence of mobile communications and Internet is the ultimate goal of third generation (3G) systems and will enable the fulfillment of both user requirements by means of the same platform and through new generations of access devices. There is little doubt that the wireless communications industry will drive the Internet into new markets with the introduction of 3G mobile communication systems. These developments will also influence and change the Internet itself.

Besides the urge to serve more users, new data-oriented applications with their need for higher data rates are the main driving forces behind the research, development, and now ongoing introduction of 3G mobile communication systems. A third motivation for the development of 3G systems was the goal to create a single world-wide system allowing the users to communicate everywhere and anytime using a single terminal. Unfortunately, this goal has not been reached with 3G systems. After about 15 years of research, development, and standardization process, the compromise achieved in 1999 was to adopt a so-called family of standards, comprising five different radio interfaces with the intention of establishing interfaces between the core networks to enable communications between users served by different standards. However, three of the five radio interfaces are based on W-CDMA (Wideband-Code Division Multiple Access) technology, easing the implementation of multi-mode terminals for global roaming. The family of standards concept was introduced to ensure an evolutionary path from existing 2G systems to future 3G systems, as will be further pointed out in chapter 5. This would not have been possible with only a single standard. On the way from 2G to 3G, several enhancements for 2G systems, often called 2.5G, have been or will be implemented. For example, one major improvement for GSM is GPRS (General Packet Radio Service), which allows for packet-oriented data transfer with rates up to 160 kbps. The 3G family member UMTS FDD (Universal Mobile Telecommunication System, Frequency Division Duplex) was the first to be launched in Japan in October 2001 with several experimental UMTS FDD test systems operating in Europe and Japan months before. This launch was initially planned for the second quarter of 2001 but had to be delayed due to technical problems. It is expected that in Europe the first 3G systems will start operation towards the end of 2002.

The big economic success of 2G systems, at least until the end of the year 2000, created even higher expectations for 3G systems. Because the 3G licenses were auctioned off in many countries, especially in Europe, the expected high revenues led to extremely high licensing fees. This produced some scepticism about the likely economic success of 3G systems. It was argued that today's low prices for 2G services will make it impossible to charge the higher fees for 3G services needed to cover the high license and deployment costs of 3G systems. As at the end of the year 2000 the rapid growth of the mobile communications market slowed down due to the high market penetration already achieved and this scepticism was further strengthened. In addition, critical experts pointed out that 3G systems will bring only limited enhancements compared to 2.5G systems like GPRS. Furthermore, the high requirements of the 3G radio interfaces, especially for the W-CDMA-based ones, makes the development of competitive terminals and base stations a truly challenging task.

Despite this list of possible problems with the introduction of IMT-2000 (International Mobile Telecommunications, the official acronym for 3G systems), we have no doubt that 3G systems and services will be widely deployed and accepted by the users. There is no alternative to 3G systems for satisfying the certainly existing and steadily rising demand for accessing any type of information on the move, since 3G extends most of the applications and services now provided to fixed network customers to mobile users at equivalent or even better performance and Quality of Service (QoS) levels. Of course

WLANs (Wireless Local Area Networks) can offer the same at much higher data rates (up to 54 Mbps in the near future) with probably much cheaper equipment. However, these WLAN systems can only provide very limited geographic coverage, making cellular 3G systems the only alternative for real wide area coverage. If proper inter-working with 3G networks is ensured, WLANs can enhance their wireless access capabilities, e.g., in high traffic hot spot areas, thereby increasing the overall network capacity. Furthermore, we think the introduction of 2.5G systems, with their much higher data rates compared to 2G systems, should not be viewed as a competitor but as a major step toward 3G instead. If people get used to technologies like GPRS, and well-tailored applications and popular services are available, the demand for "more", in terms of applications, contents, and data rate will be created automatically. This will lead to a gradual increase in the market penetration of 3G systems. From a user perspective, multi-mode and multi-band devices will be the first step in this transition from 2G to 3G. Maybe the adoption of 3G will be slower than initially assumed, but we are confident that after several years 3G systems will be as commonly used as 2G systems are today.

The key enabler for the 3G consumer mass market is the packaging of applications and presentation of content within an easy-to-use terminal device. The users do not care about the technology behind their terminals but they do care about services and costs. 3G systems are being designed to give a high degree of freedom for creating new and advanced applications and services for the information society. A wide variety of them are being and will be developed in many areas such as communication, Internet access, videoconferencing, distance education, e-commerce and financial services, location-based services, personal navigation, personal health, security, remote monitoring and controlling, social networks, and leisure. Even classes of services previously undefined can be developed offering new business opportunities for all segments of the wireless industry (equipment and IC manufacturers, network operators, contents and application providers). Most of these advanced wireless services will only be possible with 3G systems because different services have different QoS requirements that can only be satisfied by 3G networks.

The future of the wireless information society will not be driven by technology. Users will decide what are the most appropriate services for their needs. The technology only defines the boundaries of what can be achieved. Users will need a device of their choice that will adapt to the programmed profile and gives access to the possible services. The device type will be determined by the application and present the content made available by the service. The next generation of mobile terminals is expected to be capable of dealing with multimedia content (voice, data, text, image, and slow-scan video) and based on a combination of functions seen in the notebook, the palm-sized hand-held and the mobile phone. This means that for 3G services and applications, a large variety of mobile terminals targeted at various market segments will emerge, with voice as still one of the main features.

2. Fundamentals

This chapter presents some of the fundamentals of digital mobile communications on which we will rely in the remainder of this book. As there are numerous textbooks which cover these topics in detail we will restrict ourselves to a very condensed presentation of the material. We will go into detail only for certain topics that are vital to the understanding of the material presented in subsequent chapters. Frequent references to the literature will be made to provide the reader with in-depth treatment of the various topics.

2.1 Digital Modulation

In today's mobile radio systems, solely digital modulation techniques are applied. Therefore, we do not cover the analog amplitude modulation (AM) and frequency modulation (FM) techniques, but refer the interested reader to the literature, e.g., to [3, 4, 5]. Subjects common to all modulation formats will be presented before describing the different digital modulation techniques.

2.1.1 Equivalent Baseband Representation

In communications engineering we usually have to deal with passband signals, for which the most generic and basic definition is that the Fourier transform $S(f)$ of the signal $s(t)$ fulfills the condition

$$S(0) = 0 \,, \tag{2.1}$$

which implies that the signal contains no DC component. This weak condition is sufficient for the theory of equivalent baseband representation. In real applications and especially in mobile radio, the spectra of the signals we have to deal with are usually concentrated around some carrier frequency f_0 which is much higher than the bandwidth of the signal itself.

We start by considering different representations of the passband signal. As we deal with signals in realizable technical systems, these passband signals are real valued.

$$s(t) = s_I(t) \cos(2\pi f_0 t) - s_Q(t) \sin(2\pi f_0 t) \tag{2.2}$$

$$s(t) = \Re \left\{ s_L(t) \exp(j 2\pi f_0 t) \right\} \tag{2.3}$$

$$s(t) = |s_L(t)| \cos(2\pi f_0 t + \theta(t)) \tag{2.4}$$

Here, (2.2) is referred to as the *quadrature component* or *I/Q* representation, where $s_I(t)$ and $s_Q(t)$ are the inphase (I) and the quadrature phase (Q) component, respectively of $s_L(t)$, the complex valued *equivalent baseband signal*. $s_L(t)$ is also called the *complex envelope* of the real signal $s(t)$. In the second representation (2.3) the signal $s(t)$ is the real part of the product of the complex envelope $s_L(t)$ with the complex carrier $\exp(j2\pi f_0 t)$. If we express the complex valued equivalent baseband signal $s_L(t)$ in terms of magnitude ($|s_L(t)|$) and phase ($\theta(t)$), this leads to the third representation formulated in (2.4). What remains is to define how we derive the complex valued equivalent baseband signal $s_L(t)$ from our real band-pass signal $s(t)$. We do this by introducing the so-called *analytic signal* $s_+(t)$ which contains only the positive frequencies of $s(t)$.

$$S_+(f) = \begin{cases} 2S(f) & \text{for} \quad f > 0 \\ 0 & \text{for} \quad f \leq 0 \end{cases} \tag{2.5}$$

$$S_+(f) = [1 + \text{sign}(f)]S(f) \tag{2.6}$$

Here sign(.) is the signum function. If we now shift the spectrum $S_+(f)$ of the analytic signal $s_+(t)$, which is centered at frequency f_0, down to baseband, we find the spectrum $S_L(f)$ of the equivalent baseband signal $s_L(t)$.

$$S_L(f) = S_+(f + f_0) \tag{2.7}$$

$$s_L(t) = s_+(t)\exp\{-j2\pi f_0 t\} \tag{2.8}$$

A vast amount of material on signals and their mathematical representation can be found in [3, 4, 5, 6, 7, 8, 9, 10].

2.1.2 Power Spectral Density

Random signals are widely used in communications engineering since only the random nature of the signals ensures that information is transmitted. Therefore, as the signals themselves are not known in the communication system, we can only define their statistical properties. The usual signal description in the time and frequency domain from a signal $s(t)$ and its Fourier transform $S(f)$, which is the amplitude density spectrum of $s(t)$, cannot be applied. Instead, we use the autocorrelation function $\phi_{ss}(\tau)$ of our now random signal $s(t)$, for which we assume stationarity and ergodicity,

$$\phi(\tau) = E\{s(t)s(t+\tau)\}, \tag{2.9}$$

with $E\{.\}$ denoting the expectation value. We find the representation in the frequency domain if we take the Fourier transform of the autocorrelation function.

$$\Phi(f) = \int_{-\infty}^{\infty} \phi(\tau)\exp\{-j2\pi f\tau\}d\tau \tag{2.10}$$

$\Phi(f)$ is the *power spectral density* or *PSD* representing the distribution of power as a function of frequency for the stochastic signal $s(t)$. If we take the inverse Fourier transform at $\tau = 0$ we find

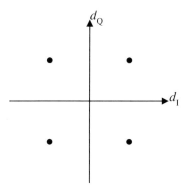

Fig. 2.1. Constellation diagram for QPSK modulation

$$\phi(0) = \left\| \int_{-\infty}^{\infty} \Phi(f) \exp\{j2\pi f\tau\} \mathrm{d}f \right\|_{\tau=0} = E\left\{ s^2(t) \right\}, \qquad (2.11)$$

which is the average power of the stochastic signal.

A detailed treatment of stochastic processes and random signals is given in [3, 4, 6, 9, 11, 12, 13].

2.1.3 Constellation Diagram

In general, linear modulated communication signals can be represented by:

$$s_\mathrm{L}(t) = \sum_{n=-\infty}^{\infty} d(n)g(t - nT). \qquad (2.12)$$

The equivalent baseband signal $s_\mathrm{L}(t)$ is the sum of time-shifted impulse responses $g(t)$ of the transmit pulse shaping filter where each is weighted with the corresponding (in general) complex data symbol $d(n)$. While $g(t)$ is a deterministic signal (since we know the impulse response of the transmit filter), the complex data symbols $d(n)$ carry the information and are the outcome of a stochastic process. The constellation diagram is the representation of the data symbols in the complex plane. A simple example is shown in Fig. 2.1, where the constellation diagram for *quadrature phase shift keying (QPSK)* is depicted. The data symbols all have the same amplitude, but each symbol can take one value out of four possible phases (see section 2.1.4). After demodulation and detection in a receiver, a decision must be made as to which data symbol was transmitted. A diagram in which the signal constellation of the actual signal is depicted, including distortions introduced by non-ideal components, gives valuable information about the signal integrity in the various stages of the transmitter and receiver. The constellation diagram thus serves as a qualitative measure of the modulation accuracy of the signal. A quantitative measure used especially in UMTS is introduced in chapter 6.8.2.

2.1.4 Linear Modulation Techniques

As already known from the previous section, we can represent a linear modulated digital communications signal according to (2.12), that is

$$s_L(t) = \sum_{n=-\infty}^{\infty} d(n)g(t - nT) \tag{2.13}$$

for the equivalent baseband representation, while for the passband representation we have

$$s(t) = \left[\sum_{n=-\infty}^{\infty} d_I(n)g(t - nT) \right] \cos(2\pi f_0 t)$$

$$- \left[\sum_{n=-\infty}^{\infty} d_Q(n)g(t - nT) \right] \sin(2\pi f_0 t). \tag{2.14}$$

Again, the equivalent baseband signal is the sum of time-shifted impulse responses $g(t)$ of the transmit pulse shaping filter which are weighted with the (in general) complex data symbols $d(n) = d_I(n) + jd_Q(n)$. For linear modulation techniques, the amplitude of the transmitted signal varies linearly with the data symbols $d(n)$ [14] and, therefore, the superposition principle with regard to the mapping of the data symbols onto the transmit signal is valid [6]. In general, linear modulation techniques are bandwidth efficient but they suffer from strong amplitude variations in the transmitted signal resulting in the need for highly linear building blocks in both the transmitter and receiver.

While $g(t)$ is a deterministic signal (since we know the impulse response of the transmit filter) the complex data symbols $d(n)$ carry the information and are the outcome of a stochastic process. The transmit filter impulse response can be chosen independently from the actual applied modulation type. It is dependent on the actual implementation and of course determines together with the modulation format the spectral properties of the transmitted signal (see also section 2.1.6). A widely used transmit filter in mobile communications is the *root-raised cosine* type filter. To derive this filter we start with the well-known *raised cosine* filter. Its impulse response, written here as a non-causal function, and frequency domain transfer function are given by

$$c(t) = \frac{1}{T} \cdot \frac{\sin\left(\pi \frac{t}{T}\right)}{\pi \frac{t}{T}} \cdot \frac{\cos\left(r\pi \frac{t}{T}\right)}{1 - \left(2r \frac{t}{T}\right)^2} \tag{2.15}$$

and

$$C(f) = \begin{cases} 1 & \text{for } |f| < \frac{1-r}{2T} \\ \frac{1}{2}\left[1 + \cos\left[\frac{\pi}{2r}\left(2fT - (1 - r)\right)\right]\right] & \text{for } \frac{1-r}{2T} \leq |f| \leq \frac{1+r}{2T} \\ 0 & \text{for } |f| > \frac{1+r}{2T} \end{cases} \tag{2.16}$$

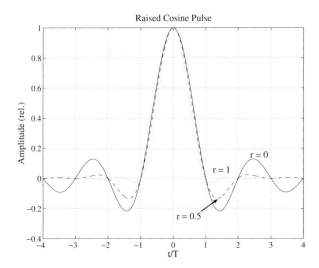

Fig. 2.2. Impulse response of a raised cosine filter with different values of r

The parameter r is the so-called *roll-off factor* describing the width of the transition from passband to stopband, where r ranges from 0 to 1. The transition itself has the shape of an elevated cosine resulting in the name "raised cosine" filter. In the case of $r = 0$ the raised cosine filter becomes an ideal low-pass filter with a rectangular frequency response and a cut-off frequency of $f_c = 1/2T$. If $r = 1$, the filter frequency response is a pure cosine and the cut-off frequency doubles compared to the case with $r = 0$, i.e., $f_c = 1/T$. The impulse response and spectrum of a raised cosine filter are displayed in Figs. 2.2 and 2.3 for different values of r. The most important feature of the raised cosine filter is that it *fulfills Nyquist's first criterion*. If this criterion is satisfied, intersymbol interference (ISI) free transmission is guaranteed. The condition to be fulfilled in the time domain is simple. The impulse response $c(t)$ of the entire system must have zeros at every sample point nT except at the main peak of the impulse response (which would be $n = 0$ in the case of the non-causal representation of the impulse response of the raised-cosine filter, for example). If we transfer this condition to the frequency domain we end up with

$$\sum_{n=-\infty}^{+\infty} C(f - \frac{n}{T}) = \text{const.} \tag{2.17}$$

Therefore, if we take the spectrum $C(f)$ of the whole transmission system, repeat it shifted with the symbol rate $1/T$ and sum up, the result must be a constant.

For the case of an ideal transmission channel that introduces no distortion, we would like to split the desired raised-cosine frequency response evenly be-

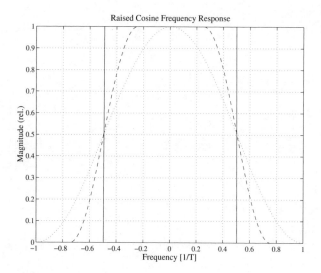

Fig. 2.3. Magnitude of the frequency transfer function of a raised cosine filter with different values of r

tween transmit $(G(f))$ and receive filter $(R(f))$ and match the receive filter to the transmit filter as well. From the matched filter condition in the frequency domain $R(f) = G^*(f)$ we derive $C(f) = G(f)R(f) = |G(f)|^2$. Thus, if we have equal transmit and receive filters, both with frequency transfer functions equal to the square-root of the raised-cosine filter, not only the first Nyquist criterion is fulfilled (if no other filtering occurs in the system), but also the receive filter is matched to the transmit filter, resulting in maximum signal-to-noise ratio (SNR) at the sampled output of the matched filter.

The resulting filter, widely used in mobile communications, is the *root-raised cosine* type filter with its impulse response $g(t)$ [10] written here as a non-causal function.

$$g(t) = \frac{\sin\left[\pi(1-r)\frac{t}{T}\right] + 4r\frac{t}{T}\cos\left[\pi(1+r)\frac{t}{T}\right]}{\pi\frac{t}{T}\left[1-\left(4r\frac{t}{T}\right)^2\right]} \tag{2.18}$$

The associated frequency response can be formulated as

$$G(f) = \begin{cases} 1 & \text{for } |f| < \frac{1-r}{2T} \\ \cos\left[\frac{\pi}{4r}\left(2fT-(1-r)\right)\right] & \text{for } \frac{1-r}{2T} \le |f| \le \frac{1+r}{2T} \\ 0 & \text{for } |f| > \frac{1+r}{2T} \end{cases} \tag{2.19}$$

In Figs. 2.4 and 2.5 the impulse response and the magnitude of the frequency transfer function of a root-raised cosine filter with $r = 0.22$ is depicted.

Coming back to the mathematical representation of linear modulated signals in (2.12) and (2.14), we find the generic transmitter architecture depicted

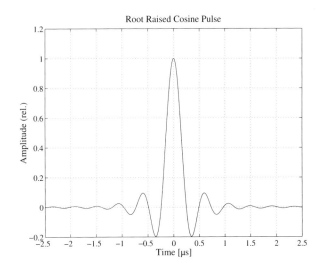

Fig. 2.4. Impulse response of a root-raised cosine filter with $r = 0.22$

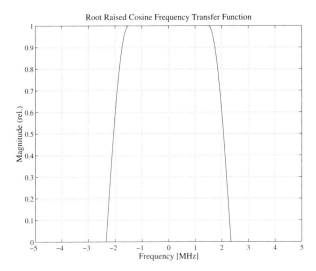

Fig. 2.5. Magnitude of the frequency transfer function of a root-raised cosine filter with $r = 0.22$

in Fig. 2.6. Depending on the actual modulation format, the serial stream of bits is converted to a stream of m parallel bits. Each m-bit-wide data word is subsequently mapped onto one valid point in the constellation diagram, yielding the real $(d_I(n))$ and imaginary $(d_Q(n))$ part of the data symbol $d(n)$. In both the I and Q paths the pulse-shaping filter follows. The outputs of the filters are multiplied with the carrier frequency $(\cos(2\pi f_0 t)$ in the I path and

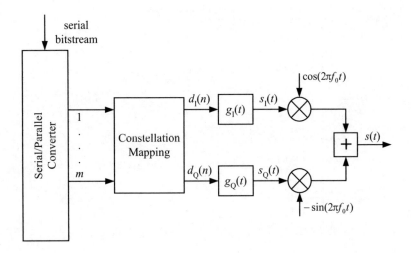

Fig. 2.6. Generic transmitter architecture for linear digital modulated signals

$-\sin(2\pi f_0 t)$ in the Q path) to translate the baseband signal to the carrier frequency f_0. Finally, we have to sum both paths to yield the real-valued band-pass signal for transmission. This generic architecture can be applied to any linear modulation format. Only the serial-to-parallel conversion and the mapping algorithm are specific for each type of modulation.

In the following, some of the most common modulation formats together with their properties are described.

Binary Phase Shift Keying. In the case of Binary Phase Shift Keying (BPSK), each symbol has two different states; therefore, with one symbol we transmit just one bit. The two different states are distinguished by their phase which can be either $0°$ or $180°$, resulting in $d(n) \in \{\pm 1\}$. Thus, the serial-to-parallel converter and the Q path in our generic transmitter architecture can be omitted. The corresponding constellation diagram and transmit signals are shown in Figs. 2.7 and 2.8.

Quadrature Phase Shift Keying. With QPSK, two bits per symbol are transmitted. The constellation diagram and the transmit signal for QPSK are depicted in Figs. 2.9 and 2.10. From a look at the constellation diagram we can consider the QPSK modulation as the superposition of two BPSK modulated signals but with orthogonal carrier signals, namely $\cos(2\pi f_0 t)$ and $\sin(2\pi f_0 t)$. The data symbols are now complex ($d(n) \in \{\pm 1 \pm j\}$). QPSK modulation is the same as quadrature amplitude modulation (QAM). With QAM the I and Q component of the signal can each take two different amplitudes (± 1), giving a total of 4 different locations in the constellation diagram, equal to QPSK. A problem associated with QPSK is the high dynamic of the complex envelope. A commonly used measure to describe this dynamic behavior is the *crest-*

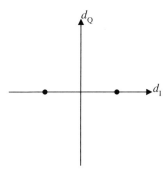

Fig. 2.7. BPSK constellation diagram

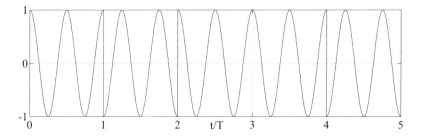

Fig. 2.8. BPSK transmit signal

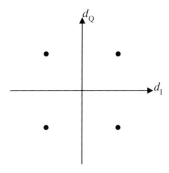

Fig. 2.9. QPSK constellation diagram

factor of a signal, which is the peak-to-average power ratio (see section 2.4.3). Due to the finite length of time to switch between states in the constellation diagram (e.g., finite time for flip-flops to transition), an amplitude ripple is introduced in the modulated signal. Therefore, a phase-shift of $\pm 180°$ (e.g., a transition from $\{+1, +1\}$ to $\{-1, -1\}$) results in a zero-crossing of the

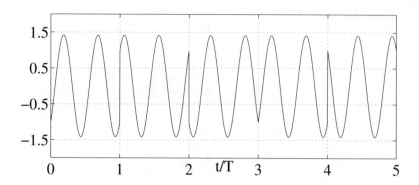

Fig. 2.10. QPSK transmit signal

envelope of the transmit signal $s(t)$. Thus, all components in the transmitter and receiver have to be highly linear to process the modulated waveforms without any significant distortions. The Offset QPSK technique alleviates this requirement to a certain extent.

Offset QPSK. The difference between QPSK and Offset QPSK is that the transitions in the I and Q branch take place at different times. The transition in the Q branch is delayed by half a symbol interval T, resulting in the expression

$$s(t) = \left[\sum_{n=-\infty}^{\infty} d_{\mathrm{I}}(n) g(t - nT) \right] \cos(2\pi f_0 t)$$

$$- \left[\sum_{n=-\infty}^{\infty} d_{\mathrm{Q}}(n) g\left(t - \frac{T}{2} - nT\right) \right] \sin(2\pi f_0 t). \tag{2.20}$$

The corresponding constellation diagram is the same as for QPSK, but the transition trajectories are different, as is shown in Fig. 2.11. The phase-shifts of $\pm 180°$ are now avoided and only $\pm 90°$ is possible, which reduces the amplitude ripple. The data symbols are the same as for QPSK ($d(n) \in \{\pm 1 \pm j\}$).

To demodulate both QPSK and Offset QPSK, the exact carrier frequency and carrier phase are needed. Recovering the carrier frequency in the receiver is not difficult, but the necessary knowledge of the phase can lead to considerable effort in the receiver's signal processing, especially in mobile communication systems, since the equalizer has to track the phase changes of the mobile radio channel (see section 2.3), which can be quite fast. Differential modulation techniques help to cope with this problem.

$\pi/4$-Differential QPSK. As an example of the Differential Phase Shift Keying (DPSK) techniques we describe the $\pi/4$-Differential Quadrature Phase Shift Keying ($\pi/4$-DQPSK) modulation [15, 16]. In principle, with differential PSK techniques, the digital information is encoded not in in the

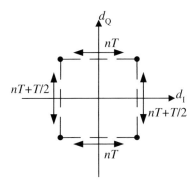

Fig. 2.11. Offset-QPSK constellation diagram

absolute phase, but in the phase difference between the actual and the previous symbol. Thus, the complex data symbol $d(n)$ can, be written as

$$d(n) = \exp\{j(\varphi(n-1) + \Delta\varphi(n))\}, \tag{2.21}$$

with $\varphi(n-1)$ being the absolute phase of the symbol at time instant $n-1$ and the phase difference $\Delta\varphi(n)$, which contains the information transmitted with symbol $d(n)$. For the case of $\pi/4$-DQPSK the phase difference takes the form

$$\Delta\varphi(n) = \frac{\pi}{2}\mu + \frac{\pi}{4}, \quad \mu = 0, 1, 2, 3. \tag{2.22}$$

Depending on the symbol value, one of the four values of μ is chosen by the encoder. The value $\pi/4$ is added to ensure that even for $\mu = 0$ a phase transition occurs. This not only aids the demodulation in the receiver but also prevents any transition through the origin of the constellation diagram, resulting in a lower crest factor. The constellation diagram is shown in Fig. 2.12. We have to note that while there are *eight* possible positions in the constellation diagram, only *four* phase transitions are possible. In Fig. 2.12 the two subsets of valid locations have different colors. If, at time instant n, our symbol takes one out of the four black locations then the symbol $d(n+1)$ has to be on one of the gray locations. The two subsets are $\pi/4$ shifted versions of each other, which results from the term $\pi/4$ added to $\Delta\varphi(n)$ in (2.22).

To achieve higher data rates the use of so-called higher order modulation schemes is possible. Typical examples are, e.g., 16 QAM up to 256 QAM for microwave radio links or 8 PSK for EDGE (Enhanced Data Rates for GSM Evolution), an advancement of GSM (see section 5.2) [17, 18]. With higher order QAM more than 2 amplitude values for I and Q component are valid (e.g., 4 in the case of 16 QAM). Similar, for 8 PSK, 8 different phase values are possible, yielding 3 data bits per transmitted symbol. Since

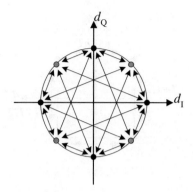

Fig. 2.12. Constellation diagram for $\pi/4$-DQPSK

the bit error probability depends on the distance between the points in the constellation diagram, the number of errors increase as the number of signal points increase. If the bit error probability is the same as with a lower order modulation scheme, the transmit signal power (which is proportional to the squared distance of the signal points from the origin of the constellation diagram) will increase.

2.1.5 Nonlinear Modulation Techniques

With nonlinear modulation techniques the superposition principle regarding the mapping of data symbols onto the transmitted waveform is violated [6]. These nonlinear modulation formats often feature a constant envelope of the transmit signal. Thus, linearity of the analog components is not a critical issue. The disadvantage of nonlinear modulation techniques is a lower spectral efficiency (see section 2.1.6) compared to linear ones.

Frequency Shift Keying. With Frequency Shift Keying (FSK), the information bits are encoded in the frequency of the transmitted waveform. The baseband representation of a M-ary FSK signal can be written as

$$s_{\mathrm{L}}(t) = \exp\left\{j2\pi d(n)\Delta Ft\right\} \cdot \mathrm{rect}\left(\frac{t}{T}\right). \tag{2.23}$$

$d(n)\Delta F$ is the frequency deviation from the carrier frequency f_0. The deviation is dependent on the value of the information symbol. The function $\mathrm{rect}(t/T)$ limits the baseband signal to a time interval of length T. Usually the frequencies used with FSK are chosen in such a way that the transmit signals for different data symbols are orthogonal. This condition can be formulated as

$$0 = \int_{-\infty}^{+\infty} s_i(t)s_j(t)\mathrm{d}t \tag{2.24}$$

$$0 = \int_{-T/2}^{+T/2} \cos\left(2\pi f_0 t + 2\pi d(i)\Delta F t\right) \cos\left(2\pi f_0 t + 2\pi d(j)\Delta F t\right)\mathrm{d}t \tag{2.25}$$

which must be fulfilled for all $i \neq j$. If we consider binary FSK with $d(n) \in \{\pm 1\}$ we derive the condition $\eta = n/2$ with $n = 1, 2, \ldots$ from the orthogonality constraint and the modulation index $\eta = 2\Delta F T$.

The problems associated with this type of modulation is that phase discontinuities can occur at a jump from one frequency to another. This results in undesired spectral broadening of the transmit signal. To ensure a continuous phase of the transmit signal we modify the FSK signal to achieve a *Continuous Phase FSK (CPFSK)* signal as follows.

$$s_{\mathrm{L}}(t) = \exp\left[j\left(2\pi\Delta F \int_0^t \sum_{n=0}^{\infty} d(n)g(\tau - nT)\mathrm{d}\tau + \varphi_0\right)\right] \tag{2.26}$$

Here $g(t)$ is a rectangular pulse of duration T, ΔF is the deviation from the carrier frequency f_0 and $d(n) \in \{1, -1\}$ for the case of binary signals. In the constellation diagram we find a trajectory of the phasor represented by (2.23). If $d(n) > 0$ the phasor rotates counterclockwise with the angular velocity $\omega = 2\pi\Delta F$ and vice versa. The angle that the phasor travels depends on the frequency deviation ΔF and the symbol duration T. The rotation angle for one symbol interval results in $|\varphi(nT) - \varphi((n-1)T)| = \pi\eta$. Depending on the actual value of η we can identify the possible signal points in the constellation diagram of which some examples are plotted in Fig. 2.13. A plot of possible phase transitions as functions of time for CPFSK signals for $\varphi_0 = 0$ is shown in Fig. 2.14. From the considerations on transmit signal amplitude variations in the case of linear modulation techniques, we note the constant envelope of FSK and CPFSK signals as a distinct advantage. This allows highly nonlinear and, therefore, highly efficient power amplifiers to be used.

Minimum Shift Keying. To limit the needed bandwidth for FSK we minimize the frequency deviation from the carrier frequency under the constraint that the orthogonality between the signals is preserved. This results in the condition $\eta = 0.5$, which leads to a rotation angle for each symbol interval of $\pm\pi/2$. Due to the minimized frequency deviation this type of modulation is called *Minimum Shift Keying (MSK)*. A plot of frequency and phase of a MSK transmit signal is shown in Fig. 2.15.

Gaussian Minimum Shift Keying. From Fig. 2.15 it can be seen that the slope of the phase trajectory of an MSK signal still can change abruptly. To smooth these sharp bends we introduce baseband filtering. The impulse response $g(t)$ of the baseband filter is changed from rectangular shape to Gaussian shape leading to *Gaussian Minimum Shift Keying (GMSK)*.

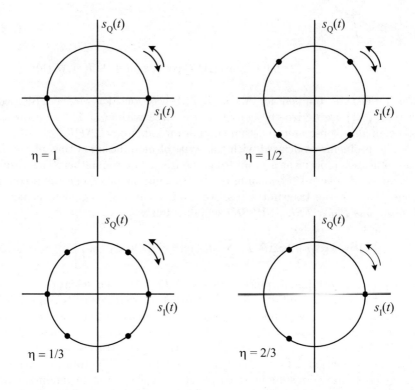

Fig. 2.13. Constellation diagram of CPFSK signals for different values of η

$$g(t) = \sqrt{\frac{2\pi}{\ln(2)}} B \exp\left(-\frac{2\pi^2 B^2}{\ln(2)} t^2\right), \qquad (2.27)$$

with the 3 dB bandwidth B. Figure 2.16 shows the now smooth phase trajectory of a GMSK signal for a time-bandwidth product of $BT = 0.3$ in comparison to the MSK phase.

2.1.6 Bandwidth Efficiency and Spectral Properties

A widely used measure for comparing modulation techniques is the bandwidth efficiency η_S, which describes the achievable bit rate per Hz. Therefore, the definition is

$$\eta_S = \frac{\text{transmission rate}}{\text{required bandwidth}}, \qquad (2.28)$$

which gives bits/s/Hz as the unit for η_S. The theoretical values for the bandwidth efficiency of different modulation techniques are listed in Table 2.1. With BPSK only one bit per clock cycle can be transmitted, resulting in $\eta_S = 1$. With higher-order modulation formats more bits per clock cycle can

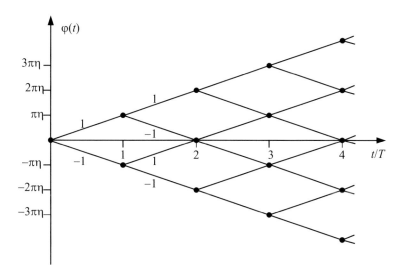

Fig. 2.14. Possible phase transitions of a CPFSK signal

be detected resulting in, e.g., $\eta_S = 2$ for QPSK or $\eta_S = 4$ for 16 QAM. These values do not represent the effect of transmit pulse shaping. Therefore, we cannot determine the actual needed bandwidth of a modulation technique directly from its theoretical bandwidth efficiency.

To compare the bandwidth occupancy of modulation techniques, the PSD is a useful tool. For linear modulation techniques we will find a simple expression to determine the PSD while for nonlinear modulation formats the computation of the PSD is either more involved or impossible, thus, one must rely on simulation. We refer to [6, 10, 20], for example, for further treatment of this topic. From section 2.1.2 we find the PSD of our transmit signal $s(t)$ to be the Fourier transform of the autocorrelation function,

Table 2.1. Theoretical values for the bandwidth efficiency of different modulation techniques

Modulation technique	η_S [bits/s/Hz]
BPSK	1
QPSK	2
Offset QPSK	2
$\pi/4$-DQPSK	2
MSK	1
GMSK	1
16 QAM	4

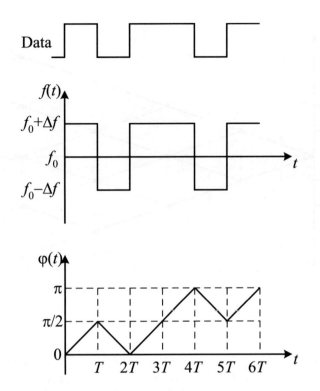

Fig. 2.15. Frequency and phase of a MSK transmit signal

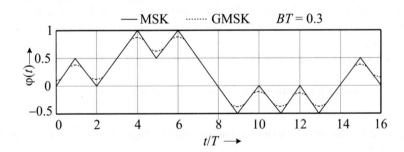

Fig. 2.16. Phase trajectories of a GMSK and a MSK signal

$$\Phi_{ss}(f) = \int_{-\infty}^{\infty} \phi_{ss}(\tau) \exp\left(-j2\pi f\tau\right) d\tau \,. \tag{2.29}$$

It is also possible to derive the PSD from the spectral properties of the random process describing our data symbols $d(n)$ and the transmit pulse shaping filter with impulse response $g(t)$ [6, 19]. First we relate the PSD

$\Phi_{ss}(f)$ of the transmit signal $s(t)$ to the PSD $\Phi_{s_L s_L}(f)$ of the equivalent baseband signal $s_L(t)$ by

$$\Phi_{ss}(f) = \frac{1}{2} \left[\Phi_{s_L s_L}(f - f_0) + \Phi_{s_L s_L}(-f - f_0) \right] . \tag{2.30}$$

We then assume that the data symbols $d(n)$ are independent of each other, identically distributed (all symbol values appear with the same probability), have a mean value μ_d and are the outcome of a random process which is wide-sense stationary, i.e., the mean value and the autocorrelation function of the process do not change with time. For this case the (discrete time) autocorrelation function $\phi_{dd}(m)$ of the random process describing the data symbols is given by

$$\phi_{dd}(m) = \begin{cases} \sigma_d^2 + \mu_d^2 & (m = 0) \\ \mu_d^2 & (m \neq 0) \end{cases} \tag{2.31}$$

where the variance of the process is denoted by σ_d^2. From this we find the PSD of the equivalent baseband signal (2.12) to be [6]

$$\Phi_{s_L s_L}(f) = \frac{\sigma_d^2}{T} |G(f)|^2 + \frac{\mu_d^2}{T^2} \sum_{m=-\infty}^{\infty} |G(f)|^2 \delta \left(f - \frac{m}{T} \right) . \tag{2.32}$$

We can distinguish two contributions to the PSD. The first term in (2.32) is a continuous spectrum depending only on the shape of the transmit filter characteristics. The second part is made up of discrete frequency components at integer multiples of $1/T$ weighted with $|G(f)|^2$ at these frequencies. Frequently, the mean value μ_d of the random process is zero (e.g., if the symbols are mapped to positions in the complex plane which are arranged symmetrical to the origin). In this case the PSD simplifies to

$$\Phi_{s_L s_L}(f) = \frac{\sigma_d^2}{T} |G(f)|^2 . \tag{2.33}$$

In the following the PSDs of different modulation techniques are shown. For all cases we have used a transmit filter with rectangular impulse response $g(t)$ and have normalized the frequency axis to the symbol duration T. Figure 2.17 shows the PSD of BPSK and QPSK. They differ in the required bandwidth by a factor of two if we assume the same bit rate. With QPSK we have half the symbol duration T as with BPSK resulting in the same bit rate. A major problem with both modulation types are the high sidelobes with the first peak being only 13 dB below the main peak. If we use a root-raised cosine transmit filter with, e.g., $\alpha = 0.22$ we get the PSD for QPSK shown in Fig. 2.18. For comparison, the PSD with rectangular transmit filter is plotted as well. With root-raised cosine filtering the spectral sidelobes are effectively suppressed but at the expense of a slightly broader main lobe. A comparison of the PSDs of MSK and QPSK is shown in Fig. 2.19. With MSK the first spectral sidelobe is 22 dB down but the width of the main lobe is

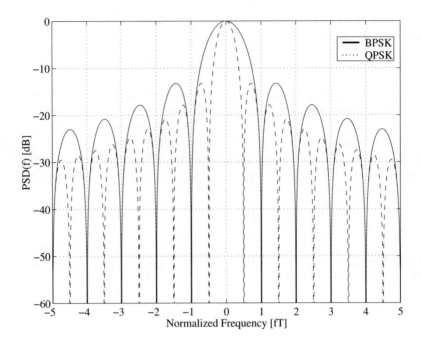

Fig. 2.17. PSD of BPSK and QPSK with transmit filter with rectangular impulse response

50% wider than with QPSK. The introduction of GMSK further lowers the sidelobes as can be seen from Fig. 2.20.

2.1.7 Bit Error Properties

Aside from the spectral properties, the bit error performance under AWGN (Additive White Gaussian Noise) conditions is a common means for comparing different modulation techniques. The following Table 2.2 summarizes the bit error probabilities (BEP) for the modulation techniques presented in the preceding sections for the case of coherent detection. Here, the complementary error function is denoted by erfc(.) and the Q-function $(Q(x))$ is the area under the tail of the Gaussian probability density function (A.1) with zero mean $(m_s = 0)$ and $\sigma^2 = 1$. The definitions of both and how they are related to each other can be found in the Appendix A in (A.4) and (A.10). The term E_b represents the signal energy per bit, while $N_0/2$ is the (two-sided) noise power density in the AWGN channel. The ratio E_b/N_0 is frequently referred to as the *signal-to-noise ratio per bit*.

The fact that BPSK and QPSK have the same BEP is only valid if both modulation types are compared at the same bit rate and with the same overall transmission energy, which can be explained as follows. QPSK can be

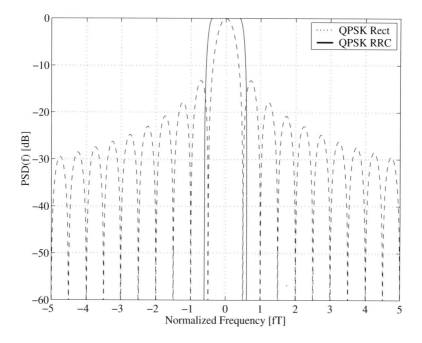

Fig. 2.18. PSD for QPSK with transmit filters with rectangular impulse response and root-raised cosine impulse response with $\alpha = 0.22$

Table 2.2. Bit error probabilities under AWGN conditions for coherent detection

Modulation	BEP
BPSK	$P_b = \frac{1}{2}\mathrm{erfc}\left(\sqrt{\frac{E_b}{N_0}}\right) = Q\left(\sqrt{\frac{2E_b}{N_0}}\right)$
QPSK	$P_b = \frac{1}{2}\mathrm{erfc}\left(\sqrt{\frac{E_b}{N_0}}\right) = Q\left(\sqrt{\frac{2E_b}{N_0}}\right)$
MSK	$P_b = \mathrm{erfc}\left(\sqrt{\frac{E_b}{N_0}}\right) = Q\left(\sqrt{\frac{2E_b}{N_0}}\right)$
GMSK[1]	$P_b = \mathrm{erfc}\left(\sqrt{\alpha\frac{E_b}{N_0}}\right) = Q\left(\sqrt{\alpha\frac{2E_b}{N_0}}\right)$

---

[1] The parameter α depends on the time-bandwidth product BT of the Gaussian pulse shaping filter. We find $\alpha \approx 0.68$ for $BT = 0.25$ and $\alpha \approx 0.85$ for $BT \to \infty$ (MSK).

viewed as two BPSK transmissions on orthogonal carriers. If we consider the same overall bit rate, then for the two BPSK systems making up the QPSK system we need only half the bit rate and, therefore, half the bandwidth of the original BPSK system. This reduces the noise power at the detector by a factor of two. But also the energy per bit is reduced by a factor of two for the two BPSK systems forming the QPSK system. So, altogether the ratio E_b/N_0 stays the same.

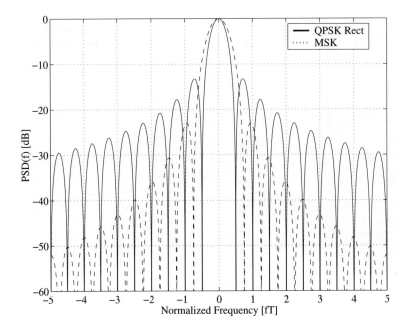

Fig. 2.19. PSD for QPSK and MSK with rectangular transmit filter impulse response

2.2 Basic Terms in Mobile Radio

In this section we describe some terms specific to mobile communication systems while others are common to communication systems in general.

2.2.1 Duplex Techniques

A duplex technique is necessary whenever a terminal is used both as transmitter and receiver. If both functions have to be provided, the transmit and receive signals must be separated. We distinguish between FDD and TDD techniques which are schematically depicted in Fig. 2.21. With FDD, transmit and receive signals are separated in frequency, i.e., each signal has its own frequency band. While this method ensures that both functions can be performed simultaneously, we have to pay for it with the need for a costly duplexer. The function of the duplexer is to separate transmit and receive path in the terminal. Since the same antenna is usually used for transmitting and receiving, the duplexer has to ensure that the transmit signal with its high power is radiated only from the antenna and not coupled into the receive path. On the other hand, all of the received signal power should be transferred from the antenna to the first low-noise amplifier (LNA) at the receiver input. Since the difference between transmit and receive power can

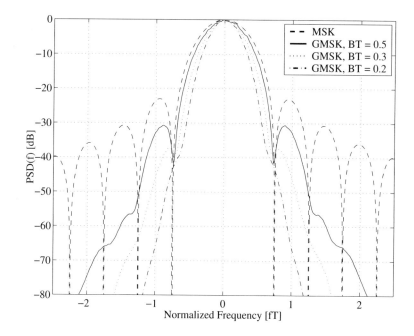

Fig. 2.20. PSD for MSK and GMSK for different values of BT

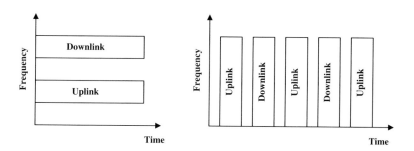

Fig. 2.21. FDD and TDD techniques for separating transmit and receive signal

be in the range of 100 dB or more, it is evident that a duplexer has to fulfill very tough requirements to ensure a satisfactory performance of the terminal.

In the second duplex technique, transmit and receive signals are separated in time (TDD). Instead of the duplexer a simple switch is sufficient in this case. The drawback is that at one time instant the terminal can only either transmit or receive. Therefore, a proper timing synchronization between the base station and all terminals served by the base station is necessary. While in today's 2G mobile radio systems mostly FDD is used, both modes are foreseen in the UMTS specification.

2.2.2 Hierarchical Cell Structure and Frequency Reuse

The introduction of the cellular concept made possible the successful deployment of mobile communication systems covering large geographical areas. The subdivision of the area to be covered into a sufficient number of cells solves two problems. First, a limited cell size also limits the transmit power, which is necessary to build mobile terminals with satisfying stand-by and talk times. Second, with the cell size the number of supported users increase. Cells that are too large result in a signal-to-interference ratio which is too low for a satisfactory operation. Each cell is served by a base station and is assigned a certain part of the frequency band. When planning the cell layout and the frequency assignment, care has to be taken, that neighboring cells operate in different frequency bands to limit the so-called inter-cell interference. Usually, the total available bandwidth is divided into a certain number of sub-bands and each cell is assigned to such a sub-band. The sub-band used in one cell can be reused in another cell if this cell is located far enough away to keep the interference level below a tolerable limit. If we group all neighboring cells with different sub-bands together we get clusters of cells. The number of cells per cluster is called the *frequency reuse factor r*, which is the same as the available number of sub-bands. The assignment of sub-bands to cells is shown in Fig. 2.22 for the case of $r = 4$. The distance between cells which use the same sub-band has to be maximized. This is easy under ideal conditions (equal cell size) but is a difficult task in a real environment. The use of ideal hexagonal cells is common because all theoretical consid-

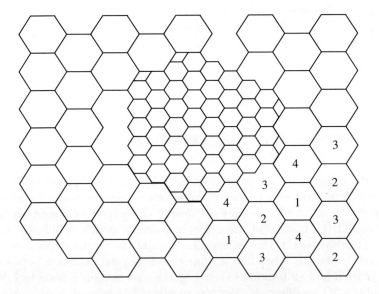

Fig. 2.22. Schematic of a hierarchical cellular net with a frequency reuse factor of $r = 4$

erations become easy and a seamless coverage of a plane area is possible. Finally, hexagonal cells closely model a circle, which is the area covered by a base station under ideal propagation conditions. The cell size is adapted to the local needs. The higher the expected traffic in a geographical area, the smaller the cell in this area must be to support the high number of users.

2.2.3 Multiple Access

In mobile communications, other than in broadcast systems, usually one transmission signal is dedicated to just one user. Therefore, we need a mechanism to separate the different user signals. We will find that a separation is possible in either frequency domain, time domain, code domain, or spatial domain. Additionally, combinations of two or more separation mechanisms are possible.

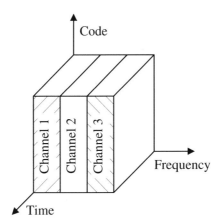

Fig. 2.23. Schematic of frequency division multiple access (FDMA)

Frequency Division Multiple Access (FDMA). The oldest type of multiple access technique is the separation of different users in the frequency domain. A unique frequency channel is assigned to each user at call setup as is shown in Fig. 2.23. To achieve a high capacity, a mobile communication system needs a high system bandwidth and a narrow channel bandwidth. Guard bands between the frequency channels must account for non-ideal filter characteristics and oscillator and component drifts, since neighboring channels must not overlap. For cellular systems using FDMA it is evident that a proper frequency planning for the hierarchical cell structure, as described in section 2.2.2, is necessary. The first analog mobile phone systems introduced in the early 1980s like AMPS (Advanced Mobile Phone Service) in

North America or the Scandinavian NMT (Nordic Mobile Telephony) relied on FDMA.

Time Division Multiple Access (TDMA). With the introduction of digital modulation techniques in mobile communication systems the separation of users in time became feasible. In a frequency channel the transmission time is divided into frames of a certain length and each frame is further subdivided into timeslots. Each timeslot in a frame is occupied by a different user (Fig. 2.24). Because of different propagation times from the mobile stations to the

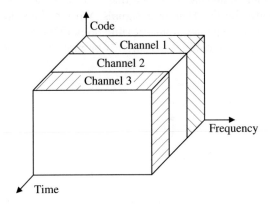

Fig. 2.24. Schematic of time division multiple access (TDMA)

base stations and because of transients associated with turning the signals on and off, guard times between the timeslots must be introduced. TDMA enhances the system capacity at the expense of additional hardware and computing effort, since we need fast switchable power amplifiers and buffering of the data stream in the transmitter and the receiver.

Code Division Multiple Access (CDMA). The direct sequence spread spectrum technique (see section 3.2) allows one to separate different users by assigning to them unique signatures, so-called codes, while their signals are transmitted in the same frequency band and during the same time (Fig. 2.25). If the receiver knows the code used in the transmitter and the codes are designed properly, the receiver can recover the transmitted user signal. For a detailed explanation of this technique, see chapter 4. The costly frequency planning for FDMA systems is not necessary in CDMA systems because they apply a frequency reuse factor of $r = 1$. A disadvantage of CDMA systems is their vulnerability to the *near–far effect* resulting in enhanced hardware effort for fast and accurate power control (see section 4.3)

Space Division Multiple Access (SDMA). If the users in a cell are located at positions with different azimuth angles they can be separated in

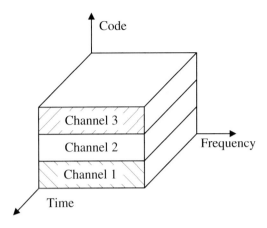

Fig. 2.25. Schematic of code division multiple access (CDMA)

the spatial domain. The most simple approach is to use sectorized antennas. This divides the cell into a fixed number of sectors where different users can only be distinguished in the spatial domain if they are in different sectors. If several antennas with the same characteristic are used instead of just one antenna, we speak of an antenna array [21]. The most simple and commonly used approach is to use a linear equally spaced array in which the antennas are distributed evenly along a straight line, all having the same orientation. The feeding points of the antennas are connected together using a combining network. The combining network can change the phase and amplitude of each antenna signal independently. The antenna pattern can be influenced if the values for the phase and amplitude manipulations are changed in the combining network. By using fixed combining networks it is possible to establish a switched beam system (Fig. 2.26a). Depending on the location of the wanted mobile station (MS) the appropriate beam is selected by a controller in the base station (BS). A more sophisticated method is to electronically control the combining network to produce several steered antenna beams. If the azimuth angles of the users differ by more than the width of the steerable antenna beam, a separation is possible (Fig. 2.26b). Up to now SDMA is not widely employed but it is expected to become a well accepted method to enhance the system capacity, while the system uses another basic multiple access system.

Hybrid Forms of Multiple Access. Usually modern mobile communication systems combine two or more multiple access schemes to achieve the highest possible capacity. Probably the most prominent system using a hybrid form of multiple access is the European GSM system [22, 23]. Figure 2.27 shows the used combination of FDMA and TDMA where each user has an assigned frequency channel and timeslot. Another example is the TDD mode

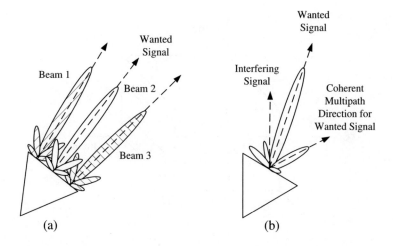

Fig. 2.26. Schematic of space division multiple access (SDMA). (a) Sectorized antennas. (b) Antenna array with electronically steerable beam

of UMTS (see chapter 7) where a combination of CDMA and TDMA is used. This means that users have their assigned timeslot but in one timeslot more than one user is active. The users in one timeslot are separated by different codes.

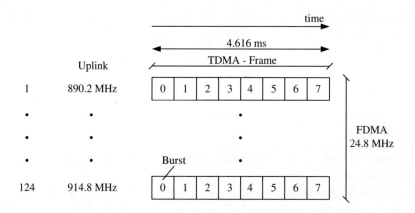

Fig. 2.27. GSM as example for a combination of FDMA and TDMA

2.3 The Mobile Radio Channel

The mobile radio channel is defined as the space between the antennas of the transmitter and receiver. As we will see, the specific conditions under which the radio wave propagation takes place have significant impact on the performance of mobile communication systems. A thorough understanding of the mechanisms in the mobile radio channel, their mathematical modeling, and their influence on the received signal characteristics is of major importance to any mobile radio communication system design. This section is intended to make the reader familiar with the features of the mobile radio channel which we will need in further parts of this book. For a more detailed description of the many aspects associated with this topic we refer the interested reader to [14, 24, 25, 26], for example. An excellent overview over the mobile radio channel, its description in terms of commonly used parameters and how to deal with the impairments introduced by the channel can be found in [27] and [28].

In principal, we have three alternatives for the description of the mobile radio channel. First we can measure the channel characteristics. This is usually very accurate but needs highly sophisticated equipment and a high number of measurements is necessary if a geographical region or a building must be completely characterized [29, 30]. A second method is to simulate the propagation in the mobile radio channel. To do so the geographical structure (e.g., a map of a city or the floor plan of a building) must be given in a digitized format since this is required by the simulation. Furthermore, propagation relevant data like roughness and complex reflection coefficients of walls and heights of buildings must be known. While simulation can give accurate results without needing measurement equipment, it can be difficult to collect the input data, and high computing effort is needed even if sophisticated algorithms are applied to reduce the necessary number of computations [31, 32, 33, 34]. A third method for characterizing the mobile radio channel is to extract some statistics, e.g., a probability density function for a certain parameter or a typical range of values for the path loss in some environment, from a large number of measurements or simulations. Given these statistics, it is possible to generate a random mobile radio channel characteristic, which, on average, has the same parameters as is found from a simulation or measurement campaign. If the statistical data are available, this method is the most convenient one providing just general channel characteristics are needed. But if, e.g., for cell planning, an accurate site specific propagation model is necessary, one would rely on simulations or measurements.

2.3.1 Overview

We can identify four distinct features of the mobile radio channel which have significant impact on the signal reaching the receiver antenna. The first property is the *multipath propagation*. The electromagnetic wave originating from the transmitter's antenna usually can arrive at the receiver via several different paths. This is due to reflection, scattering, and diffraction mechanisms, which allow for transmission paths other than the direct line-of-sight (which in many cases may not be present, e.g., because of shadowing). The multipath propagation causes fading, i.e., spatial variation of the received signal power due to constructive or destructive summing of the signals received via different paths. As a second feature we find the *time variance*. The propagation conditions can vary with time, since the mobile station can move or the obstacles causing the propagating signal to be reflected, scattered, or diffracted can change their positions. Therefore, any signal processing related to the mobile radio channel (i.e., the equalization) must be adaptive. The third property is again linked to the movement of the mobile station, namely the *Doppler spread*. This describes the spreading of a transmitted signal in the frequency domain due to the Doppler effect. While the Doppler spread is associated with time variance, it causes additional signal distortions that need to be dealt with by specialized signal processing. As a fourth feature we mention the *spatial dispersion* of signals transmitted via a mobile radio channel. Since the signals reach the receiver from different directions, the mobile radio channel characteristics vary with the azimuth angle, i.e., if we use a directional antenna at the receiver, we find different channel characteristics when we point the antenna in different directions.

In the following we will cover the aforementioned features in more detail. We will introduce key parameters to describe the properties and will show how they affect the received signal.

2.3.2 Multipath Propagation

Multipath propagation results from the propagation mechanism of electromagnetic waves. For typical mobile radio channels the free space propagation of the transmitted wave is at least partly disturbed by many obstacles. Depending on the size of the obstacle in relation to the wavelength and its position, the electromagnetic wave is reflected, scattered, or diffracted. Therefore, many signal components arriving via different paths reach the receiver antenna. We can have a line-of-sight (LOS) path if a direct path from the transmitter to the receiver is present. All other paths are so-called non-line-of-sight (NLOS) paths. Depending on their relative phases, the electromagnetic waves arriving at the receiver can either add constructively or destructively. Since the relative phases are functions of frequency and the relative propagation delay, the total received power varies strongly with the location of the receiver. These variations in the received power are called fading. We can distinguish different types of fading. First, we differentiate between *small scale* and *large scale fading*, depending on whether we perform spatial averaging on the received power or not. This distinction is shown in Fig. 2.28. Second, we can distinguish *flat* and *frequency selective fading*. This means that, depending on the bandwidth of the transmitted signal and the channel

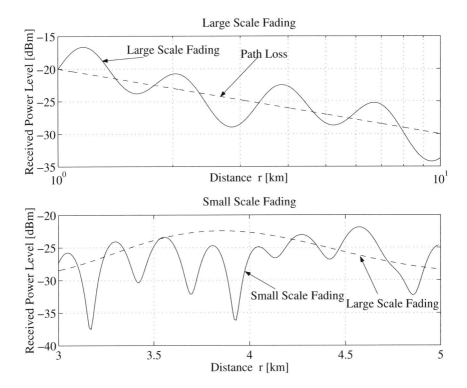

Fig. 2.28. Different types of fading

characteristics, either the whole signal is affected by the fading (flat fading), or the fading varies with frequency across the bandwidth of the signal (frequency selective fading). Another distinction is the speed of the fading effect. *Slow fading* is when the time variations are slow compared to the baseband symbol duration. Otherwise, we have *fast fading*.

2.3.3 Impulse Response Model

A simple but comprehensive model of the multipath channel is its description by means of an impulse response [35]. We consider the signal at the receivers antenna as the sum of several replicas of the transmitted signal. Each replica has its specific amplitude weighting, time delay, and phase shift. Furthermore, these three parameters for each replica can vary with time. From these considerations we find the impulse response model for the multipath channel to be

$$h(t,\tau) = \sum_{i=1}^{N(t)} z_i(t)\delta(\tau - \tau_i(t)). \tag{2.34}$$

In this equation, t is the time over which the channel impulse response may change (e.g., because of motion of the mobile station), while τ is the *multipath delay* which represents the propagation delay between transmitter and receiver for a fixed value of t. With this notation we have $N(t)$ as the number of different paths which can vary with time, $z_i(t)$ as the time-varying complex path weight for path i which accounts for the amplitude weighting and the phase shift, and $\tau_i(t)$ as the propagation delay for path i. We can now compute the received signal $y(t)$ as the convolution of the transmit signal $x(t)$ with the impulse response $h(t, \tau)$.

$$y(t) = \int_{-\infty}^{\infty} x(\tau) h(t, \tau) d\tau. \tag{2.35}$$

In the next subsection, we will define some basic parameters to describe the mobile radio channel.

2.3.4 Parameters of the Mobile Radio Channel

The following parameters are needed to distinguish between the different types of fading mentioned earlier.

For the definition of the *rms delay spread* we need the power delay profile which can be described by

$$P(\tau) = |h(\tau)|^2. \tag{2.36}$$

This describes the received power as a function of the delay τ at some fixed time instant t. The rms delay spread is defined as the square root of the second central moment of the power delay profile.

$$\tau_{\text{rms}} = \sqrt{\bar{\tau^2} - (\bar{\tau})^2} \tag{2.37}$$

Here $\bar{\tau}$ is the mean excess delay defined by

$$\bar{\tau} = \frac{\sum_i |z_i|^2 \tau_i}{\sum_i |z_i|^2} \tag{2.38}$$

and

$$\bar{\tau^2} = \frac{\sum_i |z_i|^2 \tau_i^2}{\sum_i |z_i|^2}. \tag{2.39}$$

The rms delay spread describes a kind of average value of the time dispersion of the multipath channel. Typical values for τ_{rms} are in the order of microseconds for outdoor scenarios while in indoor channels we have to deal with nanoseconds.

An analogous description of the multipath channel in the frequency domain can be found by defining the *coherence bandwidth* B_{C}. The rms delay spread and the coherence bandwidth are inversely proportional

$$B_{\text{C}} \propto \frac{1}{\tau_{\text{rms}}}. \tag{2.40}$$

The coherence bandwidth defines a frequency range over which the spectral characteristic of the mobile radio channel does not change significantly (i.e., over which the channel can be considered as "flat"). A more mathematical definition uses the correlation function of two sinusoidal signals. The value of the correlation function must be below a certain threshold value if the two sinusoids have a frequency separation greater than B_C. For this threshold we find different values in the literature. The two most frequently used values are 0.5 and 0.9 [27]. Depending on the value used we can derive different relationships between B_C and τ_{rms}. For 0.5 as a threshold value we have

$$B_C \approx \frac{1}{5\tau_{rms}} \tag{2.41}$$

while for 0.9 we get

$$B_C \approx \frac{1}{50\tau_{rms}}. \tag{2.42}$$

Both rms delay spread and coherence bandwidth are measures for the time dispersion of the channel due to *multipath effects*. However, they do not account for *time variations* due to relative motion between transmitter and receiver or due to movement of objects in the channel. Therefore, we introduce the parameters *Doppler spread* and *coherence time*. To define the Doppler spread B_D we transmit a sinusoidal signal with frequency f_0. Due to the Doppler effect because of the relative movement between transmitter and receiver the received signal has a frequency offset. Because this frequency offset is different for every propagation path, the spectrum of the received signal is not just shifted by a certain Doppler frequency but broadened and can contain frequency components in the range $f_0 - B_D \leq f \leq f_0 + B_D$ with B_D being the Doppler spread. If the transmit frequency is $2\,\mathrm{GHz}$ and the mobile station moves with a speed of $100\,\mathrm{km/h}$ B_D evaluates to $185\,\mathrm{Hz}$, while we get $926\,\mathrm{Hz}$ for a speed of $500\,\mathrm{km/h}$. The actual shape of the Doppler spectrum is different for every multipath channel. Several models are available. One of the most prominent is described by

$$S(f) \propto \frac{1}{B_D\sqrt{1 - \left(\frac{f-f_0}{B_D}\right)^2}}, \tag{2.43}$$

which is derived by using the two-dimensional dense isotropic scattering model introduced by Clarke [36].

The analogous time domain description of the Doppler spread is the *coherence time* T_C, which is inversely proportional to the Doppler spread

$$T_C \propto \frac{1}{B_D}. \tag{2.44}$$

T_C can be described as a statistical measure of the average time duration over which the mobile radio channel characteristics remain nearly the same. If the transmission of a baseband symbol lasts longer than the coherence time, the channel will change during the symbol time causing distortions at the receiver. In a similar way as for the coherence bandwidth we define T_C as the time over which the response of the channel to a sinusoidal excitation has a correlation higher than 0.5. This results in the approximate relationship [37]

$$T_C \approx \frac{9}{16\pi B_D} . \tag{2.45}$$

After a short description of the path loss we will explain the different types of fading in relation to the aforementioned parameters.

2.3.5 Path Loss

Because antennas are not able to perfectly focus electromagnetic waves, the received signal power is a function of the separation between transmit and receive antenna. As the receiver moves away from the base station, the average received power decreases. This reduction obeys an exponential law of the form

$$P_{r\,[\mathrm{dBm}]}(d) = P(d_0)_{[\mathrm{dBm}]} - 10n \log \left(\frac{d}{d_0} \right) . \tag{2.46}$$

Here $P_{r\,[\mathrm{dBm}]}(d)$ is the received power in dBm at the distance d from the base station and d_0 is a reference distance depending on the type of system and its operating frequency. In indoor systems, 1 m is an often-used value while for outdoor cellular systems d_0 can be in the range of 10 m to 1 km. n is the path loss coefficient and $P(d_0)_{[\mathrm{dBm}]}$ is the received power at the reference distance d_0. Typical values for n are in the range from 1.5 to 6 and depend on the environment (see [35] and references herein). For free space propagation and isotropic antennas, both at the transmitter and the receiver, $n = 2$, while in a street with high buildings on both sides n can be smaller than 2 due to the waveguide-like structure, and for urban scenarios without a dominant waveguide effect n is typically between 2.7 and 3.5 [14]. In rural areas the path loss is dependent on the structure of the terrain and the vegetation. To determine the path loss we have to measure the received power as a function of the distance to the base station. This received power must be determined as a spatial average over some wavelengths to get rid of the fading effects from the multipath propagation . Many measurements must be performed, all carried out on different paths traveled by the measurement equipment. After taking the average of the measurement results at every distance, a linear regression is performed to model the path loss according to (2.46) (see Fig. 2.28). The path loss coefficient n is simply the slope of the straight line representing the averaged received power.

2.3.6 Large Scale Fading

From the measurements performed to determine the path loss coefficient it becomes clear that the actual value of the received power at a given distance to the base station depends on the azimuthal position. It was found that the received power level (spatial averaged over some wavelengths) can be modeled by a *log-normal distribution*. This implies that the received power level in dB follows a Gaussian normal distribution around a mean value which is the path loss line described in the section above. The mathematical representation is the same as for the conventional Gaussian normal distribution (A.1) with the only difference being that all variables (power level x, mean value m_s, and standard deviation σ) are specified in dB. This type of power variation is called large scale fading (see Fig. 2.28). The origins for the power level variations are effects like shadowing. A simple example is if a mobile station is located in a street with LOS condition. If the mobile station moves around a street corner the situation changes to NLOS, which usually reduces the received power significantly. These power level variations occur at large spatial dimensions compared to the wavelength, leading to the term large scale fading [35].

2.3.7 Small Scale Fading

If no averaging at all is performed the received power can vary strongly if the receiver moves just a fraction of the wavelength. This so-called small scale fading is the result of the addition of several received signals with different propagation delays. Depending on their relative phases they add either destructively or constructively. Signal variations of up to 30 dB and more are possible due to this effect. The most common modeling of small scale fading is done with the Rice and Rayleigh distributions. If we assume many propagation paths with approximately equal attenuations and no LOS path we find the Rayleigh probability density function for the magnitude of the received voltage r [14].

$$p(r) = \frac{r}{\sigma^2} \exp\left(-\frac{r^2}{2\sigma^2}\right) \quad r \geq 0 \tag{2.47}$$

If a LOS path is present the Rice probability density function results [37].

$$p(r) = \frac{r}{\sigma^2} \exp\left(-\frac{r^2 + A^2}{2\sigma^2}\right) I_0\left(\frac{Ar}{\sigma^2}\right) \quad r \geq 0. \tag{2.48}$$

A denotes the peak amplitude of the LOS path and I_0 is the modified Bessel function of the first kind and zero order. Often, the so-called Rice factor that describes the ratio of received power via the LOS path to the power received via the NLOS paths is defined as

$$K = \frac{A^2}{2\sigma^2}. \tag{2.49}$$

K varies between 0 and infinity. If the Rice factor is zero there is no LOS path and we end up with the Rayleigh distribution as a special case of the Rice distribution. In Fig. 2.29 the Rice distribution for different values of K

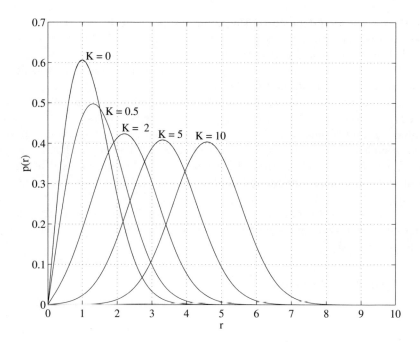

Fig. 2.29. The Rice probability density function with different Rice factors k. If $k = 0$ the Rayleigh distribution results

is plotted together with the Rayleigh distribution. Examples for the received signal power level as function in time in case of Rayleigh and Rice type fading are shown in Fig. 2.30.

Besides the description of the small scale fading by means of its amplitude probability distribution function we can distinguish different types of small scale fading depending on the relation between signal parameters and mobile radio channel parameters. The multipath effect leads to time dispersion and to fading effects which can be either *frequency selective* or *flat*. On the other hand the Doppler spread leads to frequency dispersion and either *fast* or *slow* fading. These two distortion mechanisms are independent from each other. In Table 2.3 the different types of fading together with their relations to the parameters of the transmitted signal are listed.

Flat Fading. Flat fading is caused by time dispersion effects from multipath propagation and occurs if the bandwidth of the transmitted signal is much smaller than the coherence bandwidth B_C of the mobile radio channel. In this case, the spectral characteristic of the signal is preserved but the received signal amplitude changes with time. The characteristic of the amplitude fluctuations introduced by the channel can usually be described by the Rice and Rayleigh distribution covered above. If we view the channel characteristics in the time domain we find that the symbol duration of the

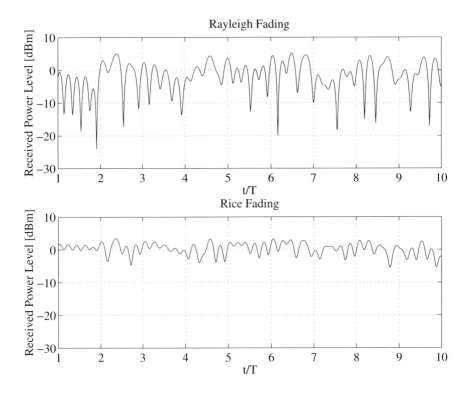

Fig. 2.30. Received signal power level for Rayleigh and Rice ($K = 10$ dB) type of fading

Table 2.3. Types of small scale fading

Small Scale Fading (based on multipath effects)	
Flat Fading	Frequency Selective Fading
signal BW < coherence BW	signal BW > coherence BW
Small Scale Fading (based on Doppler Spread)	
Fast Fading	Slow Fading
High Doppler spread	Low Doppler spread

transmit signal (the reciprocal bandwidth of the signal) is much larger than the rms delay spread. A flat fading channel can be modeled quite simply by just introducing a time-varying gain which has either a Rice or a Rayleigh distribution depending on the existence of a LOS component.

Frequency Selective Fading. Frequency selective fading is again a time dispersion effect but as opposed to flat fading the signal bandwidth is much larger than the coherence bandwidth. In the time domain, therefore, we have a rms delay spread that is much larger than the symbol duration. These effects cause different gains for different spectral components as well as *(ISI)* because multiple versions of the transmit signal (with different time delays and different gains) arrive at the receiver. An appropriate model for the frequency selective channel is the impulse response model, which has a certain number of paths each with a specific delay and a complex path coefficient. Additionally, all parameters are time-variant. Not only is the model for the channel much more complicated than for the flat fading channel, but also the necessary equalization of the channel in the receiver needs much more signal processing effort.

Fast Fading. Fast fading is an effect introduced by the Doppler spread which itself is caused by relative movement between transmitter and receiver or by moving elements in the channel. This effect is independent of time dispersion effects. Fast fading is when the coherence time T_C is much smaller than the symbol duration, i.e., the channel impulse response changes significantly during a symbol duration. The equivalent statement in the frequency domain is when the Doppler spread B_D is larger than the signal bandwidth. Due to the fact that the maximum Doppler spread is as low as 926 Hz for a transmit frequency of 2 GHz and a speed of 500 km/h, slow fading can be assumed for terrestrial mobile communication systems. Systems with low earth orbiting (LEO) satellites [38], however, feature Doppler spreads up to 50 kHz and more [26]. At low data rates (e.g., 9.6 kbit/s) fast fading is likely to occur.

Slow Fading. In slow fading channels, the change of the impulse response due to Doppler spread is much slower than the symbol duration which is equivalent to the fact that the Doppler spread is much lower than the signal bandwidth.

Because time and frequency dispersion are independent effects, a mobile radio channel can be qualified to fall into one of four different categories depending upon wether it is flat or frequency selective and wether it is fast or slow fading.

2.3.8 Spatial Dispersion

As a consequence of the multipath propagation, the radio waves transmitted from a single antenna arrive at the receiver's antenna from different directions [34]. Therefore, the description of the mobile radio channel by means of a time-variant impulse response as in (2.34) is only valid for a certain pair of transmit and receive antenna. If the receive antenna is highly directional, the measured impulse response will be different at various azimuth angles of the received antenna [39]. It can be concluded, that multipath propagation not only gives rise to time dispersion possibly resulting in ISI, but also introduces spatial dispersion. Usually certain azimuth angles are preferred in the sense that via these angles most of the signal energy is received. Each of these preferred angles show some angular spread, i.e., a certain interval in which the main part of the paths energy is concentrated. The use of antenna arrays in combination with digital signal processing, so-called *smart antennas*, makes it possible to exploit the spatial dispersion of the mobile radio channel by generating an adaptively steerable antenna beam pattern. The beam pattern is constructed in such a way that it points its main lobes into the angles with high received energy and provides deep notches in directions from which interfering signals are arriving [39].

2.3.9 Diversity Techniques

A common means to combat the impairment of the received radio signal due to fading is the introduction of *diversity*. The basic idea is simple. If several replicas of the transmitted signal are received over multiple paths which experience independent fading, then a proper combination of these replicas will not show the deep fades each path is subject to. This is because for independent paths (with respect to fading) there is only a very small probability that all paths will be in a deep fade at the same time. In fact, if p is the probability that the SNR of one path is below a certain threshold, then p^N is the probability that the SNR values on *all* N branches are below the threshold *at the same time*.

Diversity techniques rely on the statistical independence of the several replicas of the transmit signal. While it is not mandatory to have *zero* cross-correlation between the diversity branches, the cross-correlation must stay below a certain threshold otherwise the diversity scheme will not provide any significant performance gain. The most important diversity schemes are as follows [24, 40]:

Space Diversity. Two antennas are placed at the receiver's location. They must have a certain minimum distance between them, e.g., 0.5 to $0.8\,\lambda$ for the case of received signals arriving from all directions, in order to feed independent signals to the receiver input.

Angle Diversity. If the angular spread of the received signals is high, directional antennas can be used. Non-overlapping beams are necessary to guarantee uncorrelated fading on the diverse signals.

Polarization Diversity. Polarization diversity exploits the fact that signals transmitted with orthogonal polarization experience uncorrelated fading. We can either transmit and receive the same signal with two orthogonal polarization directions, or the signal is transmitted using a single polarization but received with a cross-polarized antenna. Since the scattering environment tends to depolarize the signal, the cross-polarized antenna gathers more energy than a linear polarized one.

Frequency Diversity. Frequency diversity is obtained by using several channels at different frequencies. Since the channel separation must be at least in the order of the coherence bandwidth B_C, frequency diversity is not a bandwidth efficient solution for TDMA and FDMA systems. A common type of frequency diversity is multicarrier transmission or OFDM (Orthogonal Frequency Division Multiplexing) [41, 42]. Here, several data symbols are transmitted in parallel, modulated onto many closely spaced carriers. The total transmission bandwidth is usually larger than the coherence bandwidth of the channel. Therefore, the frequency selective fading degrades some of the modulated carriers while others are mostly unaffected. Owing to the application of a powerful error correction coding scheme across the carriers, the lost data (from the carriers in the deep fades) can be recovered. Thus, multicarrier transmission does not use additional bandwidth for transmitting the same information more than once. Instead, the transmit spectrum is designed in a way that frequency selective fading can be compensated for by the system. Another type of frequency diversity is exploited by fast frequency hopping spread spectrum systems (see section 3.3). Here the carrier frequency changes several times during each symbol duration leading again to uncorrelated fading.

Time Diversity. With time diversity, identical messages are transmitted in different timeslots. Usually the same principle as with multicarrier transmission is applied but now in the time domain. The data symbols are *interleaved*, i.e., consecutive symbols of a data block are distributed over several timeslots. If one slot is lost because of fading, some error correction coding can recover the data, because only a small fraction of the data block is lost due to interleaving. Since time diversity is only effective if the diversity branches are separated by at least the coherence time, this type of diversity cannot be exploited when the mobile channel is changing only very slowly with time or not even changing at all.

Multipath Diversity. If the signal components received via different channel paths can be resolved by means of signal processing we define this as multipath diversity. This is possible in direct sequence spread spectrum systems (section 3.2) together with the so-called RAKE receiver which is explained in more detail in chapter 4.2.2.

2.3.10 Combining Techniques

After receiving our signal via the different branches using one of the above described diversity techniques, we have to combine them in the proper way. We will describe the four major combining techniques.

Selection Combining. Selection combining is the simplest technique of all. We just take the diversity branch with the highest SNR at any time instant and ignore all other branches.

Switched Combining. The difference between switched and selection combining is that with switched combining the receiver stays on the diversity branch initially chosen until the SNR drops below a certain threshold. In this case the receiver switches to another branch whose SNR exceeds the threshold.

Maximum Ratio Combining. Maximum ratio combining gives the best performance results of all combining techniques. Each branch is weighted with its complex conjugate path weight before summing up. Thus, it is necessary to know or estimate these path weights and this involves highly complex computations.

Equal Gain Combining. If the magnitudes of the path weights are difficult to estimate, we can just compensate the different phases of the diversity branches before summing them up. This results in equal gain combining, since each branch has unit gain. The performance of equal gain combining is close (within 1 dB) to maximal ratio combining.

2.4 RF Fundamentals

In this subsection we will briefly describe some some common parameters used to access the performance of radio frequency (RF) systems. We will restrict this introduction to parameters and effects which we will need later during the evaluation of the UMTS test cases and the description of already published front-end circuits.

2.4.1 Nonlinearity

In any realization of a communication system we are faced with nonlinear phenomena. The nonlinearities can be either troublesome, e.g., as in amplifiers or active filters, or are the essential property of the circuit if it is, for example, a mixer. For a system to be linear the *superposition principle* must be fulfilled, i.e., if the system's response to $x_1(t)$ and $x_2(t)$ are $y_1(t)$ and $y_2(t)$, respectively, then the response to $ax_1(t) + bx_2(t)$ is $ay_1(t) + by_2(t)$. It follows that for a nonlinear system the superposition principle does not hold. A nonlinear circuit or system usually generates frequency components not present in the input signal.

Systems, both linear or nonlinear, can be either *memoryless* or *dynamic* [43]. In the memoryless case, the output of a system does not depend on the past values of its input. In the linear case this implies the following input–output relationship.

$$y(t) = ax(t) \tag{2.50}$$

with a being a constant for a time-invariant system or a function of time for a time-variant system. A memoryless nonlinear system with a frequency independent nonlinearity can be modeled by a polynomial expression, e.g., of order 3.

$$y = a_0 + a_1 x + a_2 x^2 + a_3 x^3 \tag{2.51}$$

In a dynamic system the output depends on the past values of the input or output. For linear time-invariant systems we find

$$y(t) = x(t) * h(t). \tag{2.52}$$

Here, $h(t)$ is the impulse response of the system and $*$ denotes the convolution of $x(t)$ and $h(t)$. In the case of a linear time-variant system the output becomes

$$y(t) = x(t) * h(t, \tau), \tag{2.53}$$

with the time dependent impulse response $h(t, \tau)$. For nonlinear and dynamic systems a Volterra series approximation must be found [44] which is beyond the scope of this book.

For the following considerations we will restrict ourselves to the case of memoryless nonlinearities without phase distortions, resulting in a nonlinear model according to (2.51). An example of such a device is an amplifier. If we now excite this nonlinear amplifier with a sinusoidal signal $x(t) = \hat{x} \cos(2\pi f_0 t)$ we find the output signal $y(t)$ to be

$$\begin{aligned}
y(t) &= a_0 + a_1 \hat{x} \cos(2\pi f_0 t) + a_2 \hat{x}^2 \cos^2(2\pi f_0 t) + a_3 \hat{x}^3 \cos^3(2\pi f_0 t) \\
&= a_0 + a_1 \hat{x} \cos(2\pi f_0 t) + \frac{a_2 \hat{x}^2}{2} \left[1 + \cos(2\pi 2 f_0 t) \right] + \\
&\quad + \frac{a_3 \hat{x}^3}{2} \cos(2\pi f_0 t) \left[1 + \cos(2\pi 2 f_0 t) \right] \\
&= a_0 + \frac{a_2 \hat{x}^2}{2} + \left[a_1 \hat{x} + \frac{3 a_3 \hat{x}^3}{4} \right] \cos(2\pi f_0 t) + \\
&\quad + \frac{a_2 \hat{x}^2}{2} \cos(2\pi 2 f_0 t) + \frac{a_3 \hat{x}^3}{4} \cos(2\pi 3 f_0 t).
\end{aligned} \tag{2.54}$$

It can be seen that the output of the nonlinear amplifier contains not only the output signal of an ideal amplifier ($a_1 \hat{x} \cos(2\pi f_0 t)$) but also additional terms at f_0 and components at integer multiples of the input frequency. This is defined as harmonic distortion. The second-order nonlinear term (a_2) introduces not only a frequency component at $2 f_0$ but also an additional DC component at the output. The third-order part (a_3) of the nonlinear characteristic leads to the additional term $\frac{3 a_3 \hat{x}^3}{4}$ at frequency f_0 and a signal component with the same amplitude at the second harmonic $3 f_0$. The additional term at the fundamental frequency is an in-band distortion.

Gain Compression. An ideal linear amplifier would just multiply the input signal with the coefficient a_1. If we plot the output power versus the input power, both in dB, we find a straight line with slope 1. For real amplifiers the characteristic starts to deviate from the ideal linear regime at some specific level and finally saturates. The point at which the difference between the actual output power and the ideal linear output power is 1 dB, is called the

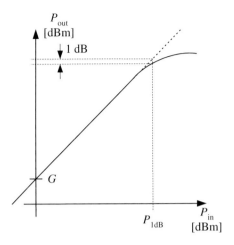

Fig. 2.31. Definition of the 1 dB compression point $P_{1\mathrm{dB}}$

1 dB compression point $P_{1\mathrm{dB}}$. A graphical representation of $P_{1\mathrm{dB}}$ is shown in Fig. 2.31. For this compressive behavior to occur, the coefficient a_3 for the third-order part of the nonlinearity must be negative. From a simple calculation we find the amplitude of the input voltage at the 1 dB compression point $\hat{x}_{1\mathrm{dB}}$ to be

$$\hat{x}_{1\mathrm{dB}} = \sqrt{0.145 \left| \frac{a_1}{a_3} \right|}. \tag{2.55}$$

As a consequence of this compressive behavior, the gain of the circuit decreases from its small signal value a_1 down to 0 if the input signal amplitude increases.

Intercept Points. In systems with (memoryless) nonlinear behavior the harmonic distortions described in the preceding subsection would constitute no problem if the harmonics fall outside the passband of the system. However, if the input of a nonlinear system contains more than one frequency, so-called *intermodulation* occurs. A common means to describe this phenomenon is the two-tone test. The input signal for the nonlinearity, which is again modeled according to (2.51), is assumed to be $x(t) = \hat{x}_1(t)\cos(2\pi f_1 t) + \hat{x}_2(t)\cos(2\pi f_2 t)$. The output of the nonlinear circuit contains the following frequency components: DC, f_1, f_2, $2f_1$, $2f_2$, $3f_1$, $3f_2$, $f_1 \pm f_2$, $2f_1 \pm f_2$, and $2f_2 \pm f_1$. If f_1 and f_2 are closely spaced, the frequency components at $2f_1 \pm f_2$ and $2f_2 \pm f_1$ are within the system passband and, therefore, introduce nonlinear distortions. For the frequency components at DC, f_1, f_2, $2f_1 \pm f_2$, and $2f_2 \pm f_1$ in the output signal we find the following components:

$$y(t) = a_0 + \frac{a_2 \hat{x}_1^2}{2} + \frac{a_2 \hat{x}_1^2}{2} +$$

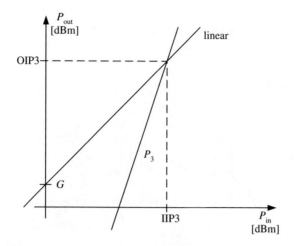

Fig. 2.32. Definition of IP3

$$\left[a_1 \hat{x}_1 + \frac{3a_3 \hat{x}_1^3}{4} + \frac{3a_3 \hat{x}_1 \hat{x}_2^2}{2} \right] \cos(2\pi f_1 t) +$$

$$\left[a_1 \hat{x}_2 + \frac{3a_3 \hat{x}_2^3}{4} + \frac{3a_3 \hat{x}_2 \hat{x}_1^2}{2} \right] \cos(2\pi f_2 t) +$$

$$\frac{3a_3 \hat{x}_1^2 \hat{x}_2}{4} \left[\cos\left(2\pi \left(2f_1 + f_2\right) t\right) + \cos\left(2\pi \left(2f_1 - f_2\right) t\right) \right] +$$

$$\frac{3a_3 \hat{x}_2^2 \hat{x}_1}{4} \left[\cos\left(2\pi \left(2f_2 + f_1\right) t\right) + \cos\left(2\pi \left(2f_2 - f_1\right) t\right) \right] +$$

$$\frac{a_2 \hat{x}_1^2}{2} \cos\left(2\pi 2 f_1 t\right) + \frac{a_2 \hat{x}_2^2}{2} \cos\left(2\pi 2 f_2 t\right) + \cdots . \qquad (2.56)$$

A common quantity to describe nonlinear distortions of building blocks like amplifiers or mixers are the so-called intercept points of third and second order, IP3 and IP2. Higher order intercept points (IPn for a nth order intercept point) can be defined in the same manner as we will show for IP3 and IP2, but they are seldom used. Only in rare cases is the fifth-order intercept point applied. Since a_3 causes distortions at or around the fundamental frequency (in-band distortions), the related IP3 is most widely used to characterize nonlinearities. IP2 is an important parameter especially in direct-conversion receivers, because the second-order nonlinearity causes a DC offset which is added to the baseband signal (see section 8.1.2).

The definition of the intercept points can easily be visualized by plotting the input-to-output power relations, as is shown in Fig. 2.32 for the case of IP3. This definition is independent of the type of nonlinearity and of the type of mathematical model used to represent it. The straight line labeled 'linear' represents the output power for the ideal linear case ($a_2 = a_3 = 0$) if both

P_{in} and P_{out} are plotted in dBm. The slope of the 'linear' line is equal to 1. For the determination of the intercept points we neglect the compressive behavior. The second straight line indicated by 'P_3' in Fig. 2.32 shows the power of the third-order intermodulation product at $2f_1 - f_2$ and $2f_2 - f_1$, which have the amplitudes $\frac{3}{4}a_3\hat{x}_1^2\hat{x}_2$ and $\frac{3}{4}a_3\hat{x}_2^2\hat{x}_1$ according to (2.56). If we assume $\hat{x}_1 = \hat{x}_2 = \hat{x}$, these terms are proportional to $\hat{x}^3$ and the associated line has a gradient of 3 in a logarithmic scale. At the point where both lines cross each other the third-order intercept point (IP3) is found. To be precise, we must always specify if we are talking about the third-order input referred intercept point (IIP3) or the third-order output referred intercept point (OIP3). Both are linked via the linear gain G (in dB) of the device

$$\text{OIP3} = \text{IIP3} + G.\tag{2.57}$$

From Fig. 2.32 we find after some simple algebraic manipulation [45]

$$P_3 = 3P_{in} + G - 2\text{IIP3}.\tag{2.58}$$

The higher IP3 (or any other intercept point) the lower the associated distortion product. If we have two input signals, $\hat{x}_1$ and $\hat{x}_2$ at the input of a nonlinear building block we find the power of the third-order intermodulation product to be

$$P_3 = 2P_1 + P_2 + G - 2\text{IIP3}\tag{2.59}$$

with P_1 and P_2 being the input powers from signal 1 and 2, respectively.

The definition of IP2 is similar to the definition of IP3. As in Fig. 2.32 we now plot the power of the second-order intermodulation products at $2f_1$, $2f_2$, or DC with amplitudes $\frac{1}{2}a_2\hat{x}_1^2$ or $\frac{1}{2}a_2\hat{x}_2^2$, respectively. Again, the crossing of both lines defines IP2. We note, that the distortion power coming from the second order component of the nonlinear characteristic occurs outside the frequency range of the input signal (both at double frequency as well as at DC). We can again find a relationship between the distortion power, now named P_2, the input power P_{in}, the gain G, and IIP2.

$$P_2 = 2P_{in} + G - \text{IIP2}\tag{2.60}$$

The *cascaded* IP of a system consisting of several building blocks, each with a different IP, can be calculated as follows. First, all IPs have to be transferred to the input of the system by subtracting gains and adding insertion losses (in dB). Second, the IPs have to be converted to powers. Third, if we can assume the distortion products to be uncorrelated, the *cascaded* IP (in mW) can be computed according to [46]

$$\text{IP}_{[mW]} = \cfrac{1}{\cfrac{1}{\text{IP}_{1,mW}} + \cfrac{1}{\text{IP}_{2,mW}} + \cdots + \cfrac{1}{\text{IP}_{n,mW}}}.\tag{2.61}$$

Here, $\text{IP}_{i,mW}$ is the IP of the ith stage in mW. Finally, the cascaded IP has to be converted from mW to dBm.

Cross Modulation. Third-order nonlinearity causes another disturbing effect if two signals, of which at least one contains an amplitude modulation component, are present at the input of the nonlinear circuit. We assume the input signal to be $x(t) = \hat{x}_1(t)\cos(2\pi f_1 t) + \hat{x}_2(t)\cos(2\pi f_2 t)$. For the general case, both $\hat{x}_1(t)$ and $\hat{x}_2(t)$ are assumed to contain an AM component (e.g., a linear digital modulated signal like QPSK). The output becomes

$$y(t) = \left[a_1\hat{x}_1(t) + \frac{3}{4}a_3\hat{x}_1^3(t) + \frac{3}{2}a_3\hat{x}_1(t)\hat{x}_2^2(t) \right]\cos(2\pi f_1 t) + \cdots, \quad (2.62)$$

and reveals that the amplitude modulation of the signal $\hat{x}_2(t)$ has been 'transferred' onto the signal $\hat{x}_1(t)$. This effect is called *cross modulation*. Of course, the same transfer of the amplitude modulation occurs if $\hat{x}_1(t)$ is a pure sinusoidal signal. Cross modulation is an important issue especially in CDMA systems, since receiver and transmitter operate simultaneously. The strong transmit signal $(\hat{x}_2(t))$ is only suppressed by the duplex filter before reaching the input of the LNA as an interferer. This can corrupt the received signal $(\hat{x}_1(t))$ directly by adding the term $\frac{3}{2}a_3\hat{x}_1\hat{x}_2^2$ to the signal at f_1 via the third-order nonlinearity of the LNA. A second deterioration occurs if a strong signal in the adjacent channel $(\hat{x}_1(t))$ is present at the input of the LNA (with third-order nonlinearity). Since the last term in the brackets in (2.62) contains the component $\hat{x}_2^2(t)$, the width of the spectrum of this term is two times the bandwidth of $\hat{x}_2(t)$ plus the the bandwidth of $\hat{x}_1(t)$. Therefore, the transfer of the AM onto the adjacent channel signal $\hat{x}_1(t)$ broadens the spectrum and signal power 'leaks' into the desired signal bandwidth [47].

Blocking. If the input signal of a circuit with compressive characteristic is the sum of a weak desired signal and a strong interferer, so-called *blocking* or *desensitization* occurs [48]. This describes the reduction of the 'average' gain of the circuit for the desired input signal due to the high amplitude of the strong interferer. If we take $x(t) = \hat{x}_1\cos(2\pi f_1 t) + \hat{x}_2\cos(2\pi f_2 t)$ the output is

$$y(t) = \left[a_1\hat{x}_1 + \frac{3}{4}a_3\hat{x}_1^3 + \frac{2}{3}a_3\hat{x}_1\hat{x}_2^2 \right]\cos(2\pi f_1 t) + \cdots. \quad (2.63)$$

If $\hat{x}_1 \ll \hat{x}_2$ this reduces to

$$y(t) = \left[a_1 + \frac{2}{3}a_3\hat{x}_2^2 \right]\hat{x}_1\cos(2\pi f_1 t) + \cdots. \quad (2.64)$$

Since $a_3 < 0$ the gain for the desired signal $\left[a_1 + \frac{2}{3}a_3\hat{x}_2^2\right]$ is a decreasing function of $\hat{x}_2$. If $\hat{x}_2$ is sufficiently large the gain tends to zero.

2.4.2 Noise Figure

A loose definition of noise can be any random interference which is not related to the wanted signal [43]. The most prominent noise sources in communication systems are thermal noise, generated by all types of resistive elements, shot noise and flicker noise from active devices (transistors), and noise received from the channel. The received noise is composed of thermal noise generated in the antenna or the transmission line and so-called "man-made" noise (e.g., any type of emissions from electronic appliances which fall into the receive band, cosmic background noise) picked up on the channel.

A common means to characterize the contamination of the wanted signal with any type of noise is the signal-to-noise ratio. It is defined as the ratio of the signal power to the total noise power. In analog transceiver front-end design, the building blocks are usually characterized with respect to their noise behavior by means of the noise figure (NF)

$$\mathrm{NF} = 10 \log \left(\frac{\mathrm{SNR_{in}}}{\mathrm{SNR_{out}}} \right) . \tag{2.65}$$

Here, $\mathrm{SNR_{in}}$ and $\mathrm{SNR_{out}}$ are the signal-to-noise ratios of the building block to be characterized, measured at the input and output, respectively. The above definition gives the NF in dB. If the conversion to dB (taking 10 times the logarithm of the ratio) is omitted, the resulting quantity is commonly defined as 'noise factor' and denoted by F.

The NF describes the degradation of the SNR due to the building block. An ideal building block which adds no noise to the signal has a NF of 0 dB (or a noise factor of 1). Real building blocks always have a NF > 0 dB. The actual noise figure of a circuit usually depends on the source and load impedance. Further details of how to compute the NF of a circuit can be found in [43] and [49], for example. Often it is important to find the overall $\mathrm{NF_{tot}}$ for a cascade of n stages. In this case, the well-known Friis formula can be applied if the noise factors are used

$$F_{\mathrm{tot}} = 1 + (F_1 - 1) + \frac{F_2 - 1}{G_{a1}} + \cdots + \frac{F_n - 1}{G_{a1} \cdots G_{an-1}} . \tag{2.66}$$

In this equation F_i is the noise factor of stage i calculated with respect to the source impedance that drives this stage. G_{ai} is the available power gain for stage i. This type of gain is defined as the ratio of available power at the output and the available power at the input of the stage [50]. Available power means that the source and load impedance are conjugate-matched to the input and output impedance of the stage, respectively.

2.4.3 Crest Factor

Although not a parameter limited to RF systems, the crest factor has considerable impact especially on the design of the RF power amplifier in the transmitter. The crest factor CF of a signal $s(t)$ is defined as the ratio of the peak amplitude to the rms value of the signal [51]

$$CF = \frac{\max\{|s(t)|\}}{\left(\lim_{\tau \to \infty} \frac{1}{2\tau} \int_{-\tau}^{\tau} |s^2(t)| \, dt \right)} . \tag{2.67}$$

In the case of a high CF the envelope of the signal shows high peaks while the rms value is low. Such a signal requires high linearity from any signal processing blocks. In the case of power amplifiers, especially high linearity results in poor efficiency, but, e.g., a mixer draws more current if higher linearity is required, too.

3. Spread Spectrum Techniques

In this chapter we will present the basic idea behind spread spectrum systems together with some of the most important issues regarding their design and implementation. We will start with a definition of spread spectrum systems and consider their advantages as well as their disadvantages, after which the different types will be described. The direct sequence spread spectrum principle will be covered into more detail including a short mathematical representation and the derivation of the optimum receiver, since UMTS relies on this technique. A following section will deal with the very important topic of synchronization and we will close with a short coverage of past and present applications together with some historical notes on the origins of spread spectrum techniques. In this chapter we of course do not intend to cover the topic of spread spectrum comprehensively. We just want to lay the theoretical background of spread spectrum for a sufficient understanding of the CDMA technology and its application within UMTS. As in the foregoing chapters we have included several references to enable the reader to cover topics of interest in more detail.

3.1 Definition

In general, every system that uses a bandwidth much wider than the bandwidth of the baseband signal can be considered as a type of spread spectrum system. This would imply that wideband FM systems can be considered as spread spectrum systems. However, throughout this book, as is common in the literature, we will use a different definition. A spread spectrum system requires that the spreading of the spectrum to be transmitted must be produced by an operation, or signal, independent of the information bearing signal [52].

C. E. Shannon expressed the basis of the spread spectrum technology by means of his channel capacity formula [53]

$$C = B \log_2 \left(1 + \frac{S}{N} \right). \tag{3.1}$$

Here the channel capacity C is the highest possible data rate which can in principle be transmitted without any error over an AWGN channel with bandwidth B and signal-to-noise ratio S/N. This is a theoretical limit which usually cannot be reached with limited effort. But 50 years of information theory has pushed today's technology within 1 dB and even less of Shannon's limit [54].

If we take a close look at (3.1) we find a linear dependency of the channel capacity on the bandwidth B, whereas C increases only with the logarithm of the signal-to-noise ratio S/N. Thus, using a higher transmission bandwidth increases the channel capacity more than by improving the S/N. Furthermore, it becomes clear that we can trade bandwidth against SNR. Using a very high bandwidth enables us to transmit over a channel with very low SNR.

We can divide the spread spectrum systems into three main categories:

- If we modulate our data signal with a digital code sequence, which is independent of the data and has a much higher clock rate than the data signal bandwidth, this is a *direct sequence spread spectrum system*.
- In a *frequency hopping spread spectrum system* the carrier frequency of the system changes in discrete steps according to a hopping pattern during transmission.
- If the carrier frequency is swept continuously over a wide bandwidth in a given pulse interval we have a *chirp spread spectrum system*.

In the following we summarize the advantages of spread spectrum systems [55]. Depending on type and application one or more of the listed advantages apply.

- The most prominent feature of spread spectrum systems, their robustness against intended and unintended interference, has already been mentioned. It will be explained in more detail in section 3.2.1.
- By using a distinct spreading characteristic for each user we achieve the capability to selectively address the users.
- If the spreading characteristics for the different users are chosen properly, their signals can be transmitted simultaneously and in the same frequency band without interfering with each other. This gives us the possibility to introduce CDMA (see section 2.2.3 and chapter 4).
- Due to the spreading of the transmit signal in the frequency domain we have low transmit power spectral densities.
- If we use proper receiver structures, the time dispersion due to multipath effects can effectively be combated.
- With spread spectrum systems a high resolution in time of arrival measurements can be achieved. This is used, e.g., in the Global Positioning System (GPS) for an accurate determination of the receiver's position.
- Privacy, because the transmission is very difficult to intercept.
- If we transmit with an SNR lower than 1 we can hide the transmit signal.

Of course we also have to deal with disadvantages. These are:

- Low bandwidth efficiency (which is only valid if we do not use CDMA).
- Accurate synchronization between the spreading process in the transmitter and the despreading process in the receiver is necessary. This usually implies high computational effort and can take a long time.
- We need fast and accurate transmit power control.
- In general the receiver structure is more complex than with conventional transmission techniques. Not only is additional signal processing needed but also the requirements for the building blocks, especially for the analog front-end are higher due to the wideband signals that must be processed.

The most commonly used quantity to describe a spread spectrum system is the *processing gain* G_P. It is defined as the ratio of transmission bandwidth B_{SS} to the bandwidth of the data signal B_D [56]

$$G_P = \frac{B_{SS}}{B_D} . \tag{3.2}$$

Another quantity used together with spread spectrum systems is the *jamming margin* M_j. The jamming margin describes how strong the interference in an AWGN channel can be without compromising the system performance. It is defined as [52, 57]

$$M_j = G_P - \left[\left(\frac{S}{N} \right)_{out} + IL \right] . \tag{3.3}$$

Here $(S/N)_{out}$ is the SNR at the spread spectrum demodulator output which is required for a sufficient performance of the system and IL represents the implementation losses.

In the following we will describe the different types of spread spectrum systems with an emphasis on the direct sequence type, since this is the technology used in UMTS.

3.2 Direct Sequence Spread Spectrum

3.2.1 Principle

Direct sequence spread spectrum (DS-SS) is the most widely used technique. The principle is quite simple. In the transmitter, the digital data signal is modulated with a code sequence $c(t)$. A schematic is shown in Fig. 3.1. The elements of the code are called chips to distinguish them from the bits in the data signal. The code sequence $c(t)$ consists of a total of L chips, each with a duration T_C. The duration of the code sequence is equal to the duration of one data bit T_D. The chip rate $R_C = T_C^{-1}$ and, therefore, the bandwidth of the signal after the modulation with the code is much higher than the data rate $R_D = T_D^{-1}$ which equals the bandwidth of the baseband signal. Examples for the corresponding signals in the time domain are plotted in Fig.

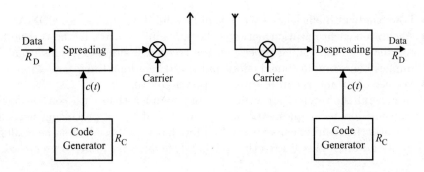

Fig. 3.1. Schematic of a direct sequence spread spectrum system

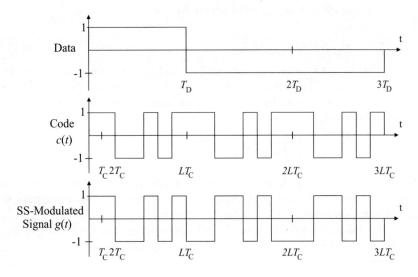

Fig. 3.2. Time domain signals in a direct sequence spread spectrum system

3.2. After translation to the carrier frequency, transmission, reception, and down-conversion to baseband, the spreading must be reversed in the receiver to recover the original data signal. By multiplying the received signal with the same code $c(t)$ the data bits are restored if and *only if the code is multiplied synchronously*. By synchronous, we imply that the code has the same phase (same starting point) as in the transmitter, the code starts exactly at the beginning of the data bit, and the code rate is exactly the same as in the transmitter. In this case all the phase changes due to chip modulation are canceled and we recover the original data signal.

The effect of the spreading is best viewed in the frequency domain (Fig. 3.3). The multiplication with the high-rate code sequence $c(t)$ broadens the transmit spectrum while at the same time the power spectral density is lowered. Again, the robustness of DS-SS systems against narrowband interferers

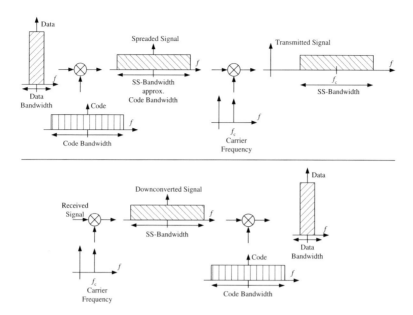

Fig. 3.3. Frequency domain representation of a direct sequence spread spectrum system

is easily explained in the frequency domain with the help of Fig. 3.4. We assume a narrowband interferer (narrowband compared to the bandwidth of the spread signal) to be added on the transmission channel. The despreading operation in the receiver, i.e., the synchronous multiplication with the code, is a despreading only for the wanted signal. The interferer, however, is spread by the multiplication with the code. Therefore, the power spectral density of the interferer is lowered while for the wanted signal it is enhanced. Depending on the processing gain, the power spectral density of the interferer on the channel can be much higher than that of the wanted signal without degrading the system performance.

For the processing gain defined in (3.2) we find for a DS-SS system

$$G_P = \frac{B_{SS}}{B_D} \approx \frac{T_D}{T_C} = \frac{LT_C}{T_C} = L \, . \tag{3.4}$$

Since the bandwidth of the spread signal is approximately the inverse of the code rate the processing gain simplifies to L, the number of chips per data bit. The processing gain describes the difference between the bit energy to interference ratio (E_b/I) and the chip energy to interference ratio (E_c/I) as

$$\frac{E_b}{I} = \frac{E_c}{I} + 10 \log(G_P) \tag{3.5}$$

if both ratios are given in dB. However, the achievable E_b/I in the receiver according to (3.5) is only valid under ideal conditions and if the interference

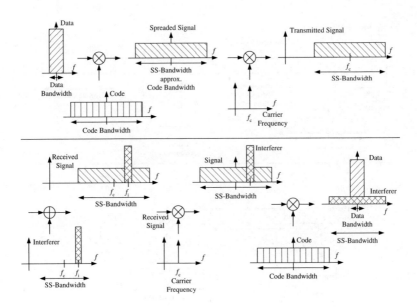

Fig. 3.4. Suppression of a narrowband interferer in a direct sequence spread spectrum system

consists merely of AWGN. Any non-ideal system component and/or interference that is correlated to the wanted signal decreases this value. This can be accounted for by an orthogonality factor OF

$$\frac{E_b}{I} = \frac{E_c}{I} + 10\log(G_\mathrm{P}) + OF \tag{3.6}$$

with OF given in dB. OF is $0\,\mathrm{dB}$ if the despreading is performed ideally and the interference is only AWGN. For non-ideal despreading OF becomes negative. Further details on OF are given in section 4.2.1 for the case that more than one user is served by the SS system.

Unlike a conventional communication system, a DS-SS system has a certain resistance against narrowband interferers depending on the processing gain, as explained above. But, with respect to noise on the transmission channel, the DS-SS system shows exactly the same bit error behavior as a non-spread spectrum system with the same modulation format. This can easily be argued, since the spreading and despreading operations on the transmitted data cancel each other out in the ideal case while they have no influence on the noise power density. The only difference to narrow band systems is that the SNR on the transmission channel is lowered by the amount of the processing gain when compared with the overall SNR.

3.2.2 Optimum Receiver

Before deriving the optimum receiver we need a mathematical description of the DS-SS signal. Following the notation for the equivalent baseband representation of linear modulated digital communication signals in (2.13)

$$s_{\mathrm{L}}(t) = \sum_{n=-\infty}^{\infty} d(n)g(t - nT_{\mathrm{D}}), \tag{3.7}$$

we have to multiply each data symbol $d(n)$ with a code sequence

$$c(t) = \sum_{i=0}^{L-1} c_i g_{\mathrm{C}}(t - T_{\mathrm{C}}). \tag{3.8}$$

Here, $g_{\mathrm{C}}(t)$ is the pulse shape of each chip. The code sequence $c(t)$ has a duration of T_{D} and is zero for $t < 0$ and $t > T_{\mathrm{D}}$ if $g_{\mathrm{C}}(t)$ is non-zero only in the interval $0 \le t \le T_{\mathrm{C}}$. There are no restrictions for the values of the chips c_i, but in most cases we will have binary codes with $c_i \in \{-1, +1\}$. We end up with the DS-SS baseband signal represented as

$$s_{\mathrm{L,DSSS}}(t) = \sum_{n=-\infty}^{\infty} d(n) \sum_{i=0}^{L-1} c_i g_{\mathrm{C}}(t - iT_{\mathrm{C}} - nT_{\mathrm{D}}). \tag{3.9}$$

The pulse shaping by means of the filter impulse response $g(t)$ is now replaced by the pulse shaping $g_{\mathrm{C}}(t)$ of each chip.

For further analysis we assume that the channel only adds white Gaussian noise $n(t)$. The received signal $r_{\mathrm{L}}(t)$ can be expressed as

$$r_{\mathrm{L}}(t) = s_{\mathrm{L}}(t) + n(t). \tag{3.10}$$

We are now looking for a receiver structure which estimates the transmitted signal "as close as possible". This can be formulated as a *least mean square* criterion

$$\int_0^{T_{\mathrm{D}}} [r_{\mathrm{L}}(t) - \hat{s}_{\mathrm{L}}(t)]^2 \, \mathrm{d}t \to \mathrm{Minimum}, \tag{3.11}$$

where the difference between received signal $r_{\mathrm{L}}(t)$ and the estimated transmit signal $\hat{s}_{\mathrm{L}}(t)$ must be minimal in a mean square sense. We find

$$\int_0^{T_{\mathrm{D}}} [r_{\mathrm{L}}(t) - \hat{s}_{\mathrm{L}}(t)]^2 \, \mathrm{d}t \to \mathrm{Minimum}$$

$$\int_0^{T_{\mathrm{D}}} r_{\mathrm{L}}^2(t)\mathrm{d}t - 2 \int_0^{T_{\mathrm{D}}} r_{\mathrm{L}}(t)\hat{s}_{\mathrm{L}}(t)\mathrm{d}t + \int_0^{T_{\mathrm{D}}} \hat{s}_{\mathrm{L}}^2(t)\mathrm{d}t \to \mathrm{Minimum}. \tag{3.12}$$

From this formulation the task to be performed in the transmitter can be derived: the received signal $r_{\mathrm{L}}(t)$ has to be correlated with all possible transmit signals. The transmit signal which gives the lowest error value is chosen as the signal which was transmitted with the highest probability. The correlation term in (3.12) must reach a maximum to minimize the error. If we just look at this correlation term for the received signal as assumed in (3.10), we have

$$\int_0^{T_D} r_L(t)\hat{s}_L(t)dt = \int_0^{T_D} s_L(t)\hat{s}_L(t)dt + \int_0^{T_D} n(t)\hat{s}_L(t)dt. \qquad (3.13)$$

Since noise $n(t)$ and signal $\hat{s}_L(t)$ are uncorrelated under the AWGN assumption, the second term on the right hand side is zero. The postulated least mean square criterion (3.11) is fulfilled if the estimated transmit signal $\hat{s}_L(t)$ equals the actual transmit signal $s_L(t)$. Therefore, we have found the optimum receiver for the detection of a known signal corrupted by additive white Gaussian noise. It can be shown that the fulfillment of the error minimization criterion is the same as maximizing the signal-to-noise ratio (SNR) of the receiver output (before the decision device).

In digital communications only a limited number of transmit symbols (e.g., $d(n) \in \{\pm 1\}$) in the case of BPSK) are possible and, therefore, a limited number of transmit signals (namely $d(n)g(t)$ with $g(t)$ as the transmit pulse shape). Thus, the optimum receiver correlates the received signal with the limited number of possible transmit signals (for BPSK with $+g(t)$ and $-g(t)$). The output of the correlators are called *decision variables*. These decision variables are compared and the signal/symbol which produces the largest one is chosen. This is essentially the same as looking for the signal which matches the received signal best. Thus, the optimum receiver bases its decision on the *Maximum-Likelihood* criterion, well known from decision theory [12]. The optimum receiver has to perform a correlation between the received signal and all possible transmitted signals/symbols. Therefore, it is called a *correlation receiver*. A schematic block diagram of this receiver for the case of binary transmit symbols $(d(n) \in \{\pm 1\})$ is shown in Fig. 3.5. After multiplying the received signal with $+g(t)$ and $-g(t)$ and integrating over one symbol duration, the sampled output values of the correlator are subtracted. If the difference is lower than zero the receiver decides for $\hat{d}(n) = -1$ and vice versa.

For SS signals, principle, theory, and performance of the correlation receiver are the same as for conventional digital modulated communication signals. The receiver now has to correlate with all possible *spread* transmit signals, i.e.,

$$\hat{s}_L(t) = \hat{d}(n) \sum_{i=0}^{L-1} c_i g_C(t - iT_C - nT_D), \qquad (3.14)$$

for the nth data symbol $\hat{d}(n)$ to be estimated.

The correlation operation can also be performed by means of a so-called *matched filter*, which has an impulse response that is the time inverse (and complex conjugate in the case of complex signals) of the transmit signal.

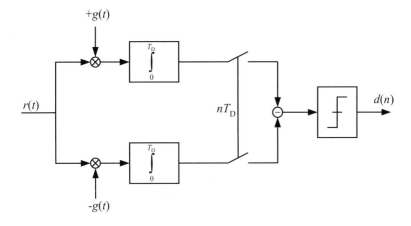

Fig. 3.5. Schematic of a correlation receiver for the case of binary signaling ($d(n) \in \{\pm 1\}$)

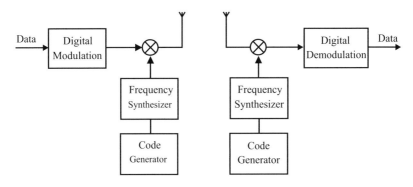

Fig. 3.6. Schematic of a frequency hopping (FH) spread spectrum system

The correlation or matched filter receiver, which is optimum for detecting a known signal after transmission over an AWGN channel, is well known in communications theory. A thorough and in-depth treatment of the associated theory can be found in many textbooks, e.g., in [6, 11, 12].

3.3 Frequency Hopping Spread Spectrum

Besides DS-SS, the frequency hopping spread spectrum (FH-SS) technique is widely known and applied. In a FH-SS system the carrier frequency of the modulated information signal is changed periodically according to a pseudo-random pattern. The schematic of such a system is shown in Fig. 3.6. The same frequency pattern must be applied in the receiver to be able to perform

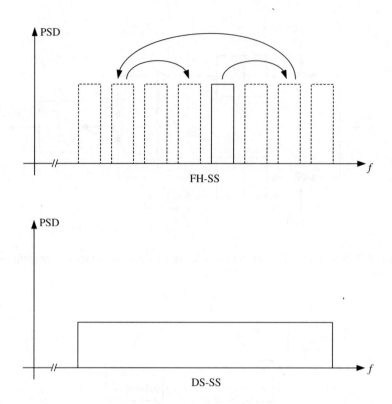

Fig. 3.7. Spectrum usage of DS- and FH-SS systems over time

the demodulation. As with DS-SS, good synchronization between transmitter and receiver is needed for a satisfactory system performance. The type of modulation applied to the carrier whose frequency is changed according to the hopping pattern is of no concern to the FH technique itself. It can be chosen depending on the system requirements.

In DS-SS systems the whole frequency band is used during the entire transmission time. This is different in FH-SS systems. Here only a small part of the systems frequency band (depending on the data modulation) is occupied but the location (in the frequency domain) of this part changes with the hopping pattern over time (see Fig. 3.7). Therefore, the total transmission power is always concentrated in the modulation bandwidth, as is the case in conventional narrowband systems. Conversely, in DS-SS systems the same transmission power is spread over the whole system bandwidth, thereby lowering the power spectral density on the transmission channel. From this it follows that FH-SS systems in operation can easily be detected. But the demodulation of a FH-SS transmission can only be performed if the hopping sequence is known in the receiver and synchronization is established, as in DS-SS systems.

For FH-SS systems we define the hopping rate R_H as the number of carrier frequency changes per time and L as the number of possible carrier frequencies. The spacing between the carrier frequencies must be large enough to prevent an overlap of spectra on the radio channel for the case that two neighboring frequencies are occupied simultaneously. Therefore, we need at least a spacing in the order of the data bandwidth B_D resulting in a total occupied transmission bandwidth of approximately $B_{SS} = LB_D$. According to the definition of the processing gain in (3.2) we find for a FH-SS system

$$G_P = \frac{B_{SS}}{B_D} \approx \frac{LB_D}{B_D} = L\,. \tag{3.15}$$

In this equation any additional spectral spreading due to the hopping process itself, i.e., the transition from one carrier frequency to another, is neglected. If these transitions occur abruptly, this assumption no longer holds.

We distinguish between two types of frequency hopping according to the relation between hopping rate R_H and data rate R_D.

- *Fast frequency hopping* (F-FH) occurs if the hopping rate R_H is (much) higher than the data rate R_D. In this case one bit or symbol is transmitted using several carrier frequencies. Without knowing the hopping sequence in the receiver demodulation is impossible. An example for the application of F-FH is the IEEE 802.11 standard for WLAN applications [58].
- *Slow frequency hopping* (S-FH) occurs if the hopping rate R_H is (much) smaller than the data rate R_D. In this case one or more bits or symbols are transmitted at one carrier frequency. S-FH is usually used to combat deep fades in a static environment (if the mobile station does not move). By hopping to a different carrier frequency after a certain number of transmitted bits the possible occurrence of deep fade affects only a limited number of bits, thereby increasing the average transmission quality. S-FH is optionally applied in the GSM system [23].

It is evident that up to L different users can transmit simultaneously without disturbing each other if the hopping sequences are designed in such a way that two users never hop to the same frequency at the same time. For this, a synchronization of the hopping sequences of all users is mandatory. If both conditions cannot be guaranteed, the number of users must be reduced to an extent that the probability that two or more users transmit at the same frequency is sufficiently low. Besides its multiple access capability the FH-SS technique also provides some resistance against multipath interference. Usually in a frequency selective mobile radio channel only some of the carrier frequencies are located in deep fades. Due to the hopping, the transmission performance is averaged over the different carriers, thereby reducing the multipath interference. Usually some error correction coding is applied in FH-SS systems. This helps in recovering the data corrupted due to strong interference at one carrier frequency.

Further properties of FH-SS systems are as follows:

- Synchronization is usually easier than in DS-SS systems since the timing requirements are not as stringent.
- FH-SS systems need highly frequency agile synthesizers in transmitter and receiver.
- The occupied frequency band must not necessarily be contiguous.
- An ultra wide bandwidth spreading is possible.
- Good near-far resistance (see section 4.3), since users with different distance to the base station usually use different frequencies.
- with FH-SS systems a coherent demodulation is difficult, because the required phase relationship is hard to maintain during hopping.

3.4 Time Hopping Spread Spectrum

In time hopping spread spectrum (TH-SS) systems the time axis is divided into frames of duration T_F. Each frame is again divided into L timeslots of duration T_S. In each frame the transmitter uses just one of the L timeslots to transmit its data packet consisting of k data bits. The position of the timeslot that is used for transmission changes from frame to frame according to some code. The processing gain is, therefore,

$$G_P = \frac{B_{SS}}{B_D} \approx \frac{T_F}{T_S} = L.$$ (3.16)

As with FH-SS in TH-SS systems, up to L different users can be active simultaneously without any mutual interference, since it is basically a TDMA system. The advantages of the TH-SS system are its implementation simplicity compared to FH-SS and the lack of the near-far problem. On the other hand the long synchronization time and the needed high timing accuracy are disadvantages.

3.5 Chirp Spread Spectrum

From radar technology the use of *chirp* signals with their wide bandwidth and the associated pulse compression technique is well known [59, 60]. Chirp signals are characterized by RF bursts whose frequency varies in a known way during each pulse period. This type of signals can also be applied to communication systems because of their inherent interference rejection capability [61, 64].

For a linear chirp waveform the instantaneous frequency f_M varies linearly with time. The waveform can be written as

$$s(t) = a(t)\cos[\Theta(t)] = a(t)\cos\left[\omega_0 t + \frac{\mu t^2}{2}\right] \tag{3.17}$$

for $-T/2 < t < T/2$, where T, $a(t)$, $f_0 = \omega_0/(2\pi)$, and μ are the chirp duration, the chirp envelope, the center frequency, and the chirp rate, respectively. The latter indicates the rate of change of the instantaneous frequency f_M. f_M and μ are defined as

$$f_M(t) = \frac{1}{2\pi}\frac{d\Theta(t)}{dt} \tag{3.18}$$

$$\mu(t) = \frac{df_M(t)}{dt} \tag{3.19}$$

$$= \frac{1}{2\pi}\frac{d^2\Theta(t)}{dt^2}. \tag{3.20}$$

In the case of a linear chirp we find a constant μ and the bandwidth of the chirp signal can be defined as the range of instantaneous frequencies, i.e.,

$$B = \frac{|\mu|}{2\pi}T. \tag{3.21}$$

We have to note that the 3-dB bandwidth of the spectrum of a chirp signal can be different from B and depends strongly on the shape of the envelope $a(t)$. If $a(t)$ is a rectangular function and if the time-bandwidth product BT is large, it can be shown that the spectrum of the chirp signal is close to a rectangular function with width B [60].

If a chirp waveform is fed into its matched filter (whose impulse response is the time inverse of the input chirp waveform and, therefore, also a chirp signal but with an opposite sign for μ) the output signal $g(t)$ of the matched filter is the autocorrelation function $\phi_{ss}(t)$. By noting that the high frequency term $(2\omega_0)$ in the result can be ignored for most cases of practical interest [60] and assuming that the envelope $a(t)$ is a rectangular function, the compressed RF pulse at the output of the matched filter becomes

$$g(t) = \sqrt{\frac{2\mu}{\pi}} \cdot \frac{\sin\left(\frac{\mu t}{2}\left(T - |t|\right)\right)}{\mu t}\cos\left(\omega_0 t\right) \tag{3.22}$$

for $-T < t < T$. This is illustrated in Fig. 3.8 for the parameters $T = 2\,\mu s$, $B = 80\,\text{MHz}$, and $f_0 = 348.8\,\text{MHz}$. The envelope of the compressed pulse has its maximum at $t = 0$, and its first zeros at $t \approx \pm 1/B$. For convenience we define the pulsewidth as $1/B$. The processing gain of chirp spread spectrum systems is defined as the product of chirp duration T and chirp bandwidth B and is called the time-bandwidth product BT.

$$G_P = \frac{B_{SS}}{B_D} \approx BT \tag{3.23}$$

Taking BT as the processing gain in the chirp spread spectrum system is common sense but can be somewhat misleading regarding the achievable data rate. This is due to the fact that the data rate can be (much) higher

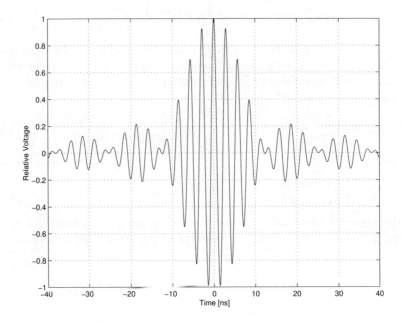

Fig. 3.8. Autocorrelation function for a linear chirp signal with the parameters $T = 2\,\mu s$, $B = 80\,MHz$, and $f_0 = 348.8\,MHz$

than $1/T$, e.g., if overlapping in time of consecutive chirp signals and/or some higher order modulation is applied. Several modulation types for chirp signals in communication systems are possible, e.g., when using binary orthogonal keying (BOK) an up-chirp is transmitted for a high-bit while a down-chirp is used for a low-bit [62]. Another possibility is the use of PSK modulation techniques on chirp signals. In this case digital phase values φ_i are added to the argument of the cosine in (3.17). It can be easily shown that these digital phase shifts, which carry the information to be transmitted, appear also in the output signal of the matched filter (3.22) [63]. The use of chirp signals for spreading the information signal aids in combating interference and multipath distortions but gives no possibility for multiple access. To realize a multiple access capability, additional coding must be introduced [64].

3.6 Synchronization

Synchronization is a major issue in communication systems. While perfect synchronization is assumed in many of the theoretical treatments the impact of synchronization on the performance of any real SS system cannot be overemphasized. In conventional systems we have to deal with, e.g., carrier, phase, frame, symbol, and clock synchronization. These points are also important in SS systems but they require an additional *code synchronization*. Code synchronization comprises not only synchronization of the code phase (to find the correct starting point for the code in the receiver) but also finding the correct code (or chip) rate. In the case of different code rates the phase differences of one chip accumulates over time and could prevent successful synchronization. Code synchronization is mandatory because otherwise any despreading would be impossible. Therefore, it is necessary to establish code synchronization *before* the beginning of the despreading. In mobile communications it is usually not possible to provide highly accurate frequency and time sources in transmitter and receiver to overcome the synchronization problems. This is not only because of the demanding requirements regarding size and power consumption especially in the mobile stations. Even if we could realize these stable sources it would not be sufficient because the usually unknown and time-varying distance between transmitter and receiver introduces an arbitrary phase shift. Additionally a possible Doppler-shift of the received signal makes sophisticated synchronization algorithms necessary.

In general we can identify the following problems in chip synchronization:

- A high timing accuracy is required. In the case of UMTS the chip duration is 260 ns. If we assume a necessary synchronization accuracy of a tenth of a chip duration, we find a necessary timing resolution of about 26 ns.
- Synchronization must be performed under low SNR conditions.
- Fast synchronization is desirable. But the longer the code the longer it will take or the more computing power is needed to achieve synchronization.

The process of synchronization is usually performed in two steps. First a coarse synchronization called *acquisition* is performed. During this operation the despreading code generated in the receiver is aligned to the spreading code of the received signal within a specified range, typically one chip period. During the following *tracking* process the remaining phase difference between the two codes is reduced as close to zero as possible. Because we know nothing about phase (of carrier and code), frame timing, clock rate, etc., at the beginning of synchronization, the acquisition is the more difficult part. While acquisition is performed once at the beginning of the communication process, tracking is a task to be performed continuously to account for any changes introduced by varying propagation conditions. Only if synchronization is lost (e.g., due to some abrupt changes in the propagation conditions the tracking loop is unable to follow) acquisition has to be performed again.

3.6.1 Acquisition

In general synchronization, i.e., acquisition *and* tracking, is based on *correlation* (see section 3.2.2). If we correlate the received signal with the code generated in the receiver, we get an autocorrelation peak from every data symbol only if the receiver code is properly synchronized.

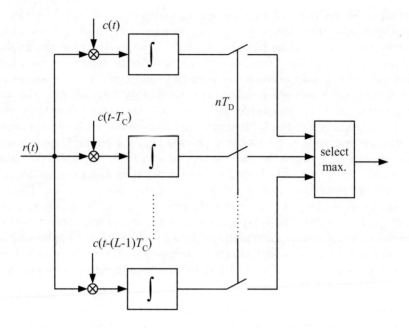

Fig. 3.9. Parallel maximum likelihood synchronization

Matched Filter Synchronization. Conceptually the simplest way of synchronization would be to use a matched filter (matched to the code we are looking for) whose output is continuously (usually at double chip rate) sampled. By searching for the maximum output of the matched filter (which is the autocorrelation peak) we are able to compute the starting phase for the code. The matched filter method is very simple and fast. A maximum of one symbol duration is needed to acquire acquisition. It is necessary for the two code rates to match accurately otherwise the autocorrelation peak will be lowered due to imperfect correlation. The matched filter synchronizer can also be used to detect one out of N codes if a bank of N matched filters is used. Owing to the associated hardware expense the application of matched filter synchronization is limited to a small number of fixed codes. A typical example would be the slot synchronization in the UMTS cell search procedure. (see section 6.6.1)

Parallel Search. Another intuitive approach to the synchronization problem is to correlate the received signal with the local code using each possible code phase. The fastest way to do this would be to use N correlators in parallel, all with the same code but different code phases, and choose the code phase which gives the maximum correlation value. This would fulfill the *maximum likelihood* criterion [65]. A schematic of such a synchronization device is shown in Fig. 3.9. Possible modifications of this structure are the use of

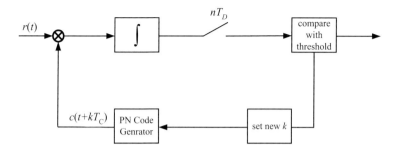

Fig. 3.10. Schematic of a synchronization device using serial search

code phases which differ in fractions of the chip duration T_C and a variable integration time for the correlation [66]. If the code phases differ by half the chip duration, a total of $2L$ correlators are necessary, but the acquired timing accuracy has been doubled. The variable integration time allows for a compromise between acquisition time and the probability of a synchronization error. The necessary hardware effort associated with such a *parallel search* for the correct code phase is usually too high to be implemented (especially if very long codes are used).

Serial Search. A serial search where the correlations for each code phase are computed one after the other drastically reduces the hardware effort. But the acquisition time is increased by the same factor in the worst case. The schematic of a serial synchronizer is depicted in Fig. 3.10. If no *a priori* information about the possible code phase is available the search can start at an arbitrarily chosen code phase and continue in a serial manner through the consecutive code phases (*straight line search*). The correlation values computed for each phase are compared to a predefined threshold. If the correlator output exceeds the threshold, the synchronization device switches from acquisition to tracking. If some information (e.g., a probability density function) on the code phase to be acquired is available, other types of search strategies like the *z-search* or the *expanding-window search* can be applied [56, 65, 67].

3.6.2 Tracking

Tracking describes the tasks of first reducing the phase offset between received and locally generated code from the acquisition range down to a small fraction of one chip duration and second to maintain this condition during the whole communication process. The two most common tracking techniques are the *Tau-Dither Loop (TDL)* and the *Delay-Locked Loop (DLL)*. Both are described in the following sections. A detailed analysis of their performance as well as other refined forms of both loops can be found in [56].

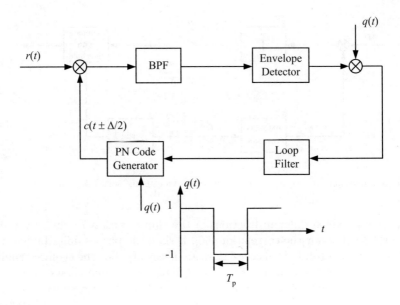

Fig. 3.11. Tau-dither code tracking loop

Tau-Dither Loop. The basic principle of tau-dither tracking is to slightly degrade the correlation (= synchronization) by a known change of the rate or phase of the locally generated code, observe the effect and use the information gained to improve the synchronization. A block diagram of a TDL is shown in Fig. 3.11 [52, 56]. The phase of the clock of the code generator is shifted periodically between two different, closely spaced states (e.g., a tenth of the chip duration). This causes the correlation between the received and the local code to change periodically, too. The output signal of the correlator is, therefore, amplitude modulated. This modulation signal is now compared in a phase detector with the rectangular signal which shifts the code phase. The phase detector delivers a signal with a DC component which has the proper amplitude and polarity to control the code clock generator. The time T_p during which the code generator uses the same phase shift must be much larger than the chip duration T_C. The principle of the TDL can be further explained if we assume the ideal triangular autocorrelation function as shown in Fig. 3.12 with the marked three different regions of phase difference between the two codes. If the two code phase differences correspond to the points labeled 1a and 1b, the phase of the AM-modulated correlator output differs by π from the phase corresponding to the point 2a and 2b. This assures the correct sign of the control voltage. The perfectly synchronized state is represented by the points 3a and 3b which cause no modulation of the correlator output.

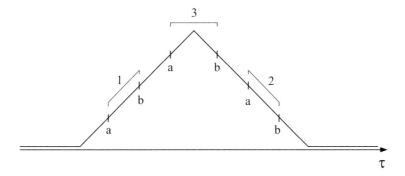

Fig. 3.12. Ideal triangular autocorrelation function with different phase shift positions

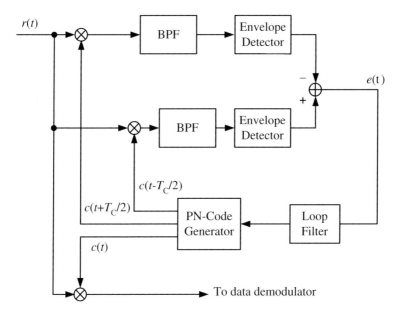

Fig. 3.13. Delay-locked loop

Delay-Locked Loop. In the DLL [52, 56] (see a schematic block diagram in Fig. 3.13) two identical codes are generated in the receiver, which are delayed in time by one chip duration with respect to each other. The received signal is correlated with both codes and the correlator outputs are subtracted. This results in an error signal which is composed of two autocorrelation functions that are delayed by one chip interval with respect to each other. The shape of the error signal is depicted in Fig. 3.14 for the case of ideal triangular autocorrelation functions. The error signal is usually filtered by a loop filter

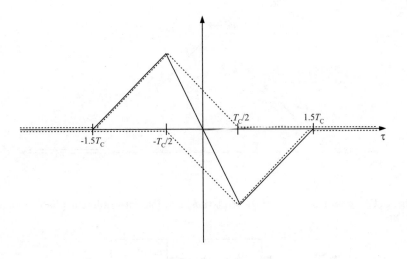

Fig. 3.14. Error signal of a delay-locked loop

whose output controls the clock of the code generator. The third output of the code generator delivers a code with a delay halfway between the two codes used for the DLL, thus yielding an exact synchronized receiver code for the SS receiver. It is essential that the two correlators in the DLL are identical, otherwise there will be a static error.

3.7 History

The origins of modern spread spectrum communications technology are found in the research efforts for robust military communication systems between about 1940 and 1960 [56, 68]. Prior to that, some fundamental technological advances in ranging and communications theory had been made which directly or indirectly contributed to the development of SS:

- The RADAR (**Ra**dio **D**etection **A**nd **R**anging) principle was discovered in the 1920s. Out of the developments in RADAR the chirp-SS technique and the matched filter concept evolved.
- One of the first patents for a delay-locked loop was filed in 1936.
- The application of the sampling theorem to communications.
- Wiener's pioneering work on probabilistic modeling of information flow in communication and control systems.
- Shannon's channel capacity [53].
- First work on correlation techniques in 1942 and 1949.

A first patent on a frequency hopping SS system was filed by Markey and Antheil in 1941 [56]. Around 1948, Mortimer Rogoff, an engineer at ITT's

Federal Telecommunication Laboratories, built a cross-correlator and developed a possibility to store noise-like signals (the so-called 'noise-wheel'), both based on photographic techniques. The 'noise-wheel' intensity modulated a light beam, thus providing the pseudo-random signal for spreading. Together with Lois deRosa, also from ITT, Rogoff published their first ideas on the DS-SS technique in 1950. In 1951, a first transmission over a distance of about 180 m took place and one year later a successful transcontinental test between Palo Alto in California and Telegraph Hill in New Jersey was performed. During the work on a DS-SS system at M.I.T. in the early 1950s, it became evident that multipath effects could severely reduce the effectiveness of SS systems. This led to the development of the RAKE receiver concept by Bob Price and Paul Green which was patented in 1956. Also in the early 1950s, pioneering work on the theory and practical generation of m-sequences for SS purposes took place. In 1962, the first FH-SS communication system, developed by Sylvania's Bayside Laboratories on Long Island, reached an operational state.

There are several reasons why the major SS developments were achieved by scientists from the USA. First, the theories of Wiener and Shannon were available in the USA before they were recognized elsewhere. Second, the SS development started after World War II when many countries suffered from considerable losses in manpower and facilities. The situation was far less severe in the USA where, with the addition of many researchers brought in mainly from Europe, scientific development progressed more rapidly. Finally, the onset of the Cold War increased the need for secure communications, especially in the USA.

This was not the case in the US which in addition brought many researchers especially from Europe to work here. Finally, the onset of the Cold War pushed the urge for secure communications especially for the US.

4. Code Division Multiple Access

As already described in section 2.2.3, a direct sequence SS system can provide *code division multiple access*. This is performed by assigning each user a different code for the spectrum spreading process. Under ideal conditions and if these codes are chosen properly, there will be negligible interference between different users despite the fact that their signals are transmitted at the *same time in the same frequency band*. Under real conditions there is of course at least some residual interference from other users which is called *multiple access interference (MAI)*. The first section of this chapter will deal with possible code families and their properties with respect to their application in CDMA systems. Afterwards we will describe the basic receiver structures with an emphasis on the RAKE receiver and briefly mention more advanced receiver concepts, so-called *multiuser detectors*. What follows is a section on interference in CDMA systems and on power control which is an important topic and closely related to MAI.

4.1 Spreading Codes

This section deals with the codes used in SS and especially in CDMA systems for the purpose of spectrum spreading and differentiation between users. We do not cover any error correction codes. Ideal spreading codes should be infinitely long to avoid any spurious signals caused by periodicity and should be totally random. In practice we need deterministic codes which appear random but are actually periodic. Today's hardware allows for very long codes so that the chosen length is basically determined by the system requirements.

4.1.1 Performance Measure

For the evaluation and comparison of codes we need some performance measurements. Because every SS or CDMA receiver is based on correlation, we can use the autocorrelation and the cross-correlation functions to judge if a code is qualified or not. We restrict ourselves to discrete correlation functions throughout this book.

From the general definition of the autocorrelation function for continuous time signals

$$\phi_{xx}(\tau) = \int_{-\infty}^{\infty} x(t)x(t+\tau)\mathrm{d}t \tag{4.1}$$

we derive a similar expression for discrete time signals. If $c_x(i)$ is the ith chip of our code x with length L (i.e., $c_x(i) = 0$ for $i < 0$ and $i \geq L$), we define the discrete time *autocorrelation function* to be

$$\underline{\phi}_{xx}(l) = \sum_{i=0}^{L-1} c_x(i)c_x(i+l) \quad \text{for} \quad 0 \leq l \leq L-1 . \tag{4.2}$$

Since the $c_x(i)$ is non-zero only for $0 \leq i < L$ we can modify (4.2) to

$$\underline{\phi}_{xx}(l) = \begin{cases} \sum_{i=0}^{L-1-l} c_x(i)c_x(i+l) & \dots \quad 0 \leq l \leq L-1 \\ 0 & \dots \quad \text{elsewhere} \end{cases} . \tag{4.3}$$

The so-called *periodic autocorrelation function* is defined as

$$\phi_{xx}(l) = \sum_{i=1}^{L} c_x(i)c_x[(i+l) \bmod L] \quad \text{for} \quad 0 \leq l \leq L-1 . \tag{4.4}$$

This is equivalent to an autocorrelation over one code length L if the code $c(i)$ is periodically repeated ($c_x(i) = c_x(i + KL)$ for every integer K). It is evident that $\phi_{xx}(l)$ is periodic with L. Therefore, this type of autocorrelation function for a time-limited signal is called the *periodic autocorrelation function* and we refer to (4.3) as the *aperiodic autocorrelation function*. To distinguish between them, the symbols for the aperiodic correlation functions are underlined throughout this text.

According to the above equations we find the continuous time cross-correlation function

$$\phi_{xy}(\tau) = \int_{-\infty}^{\infty} x(t)y(t+\tau)\mathrm{d}t , \tag{4.5}$$

the discrete time aperiodic cross-correlation function

$$\underline{\phi}_{xy}(l) = \begin{cases} \sum_{i=0}^{L-1-l} c_x(i)c_y(i+l) & \dots \quad 0 \leq l \leq L-1 \\ 0 & \dots \quad \text{elsewhere} , \end{cases} \tag{4.6}$$

and the periodic cross-correlation function

$$\phi_{xy}(l) = \sum_{i=1}^{L} c_x(i)c_y[(i+l) \bmod L] \quad \text{for} \quad 0 \leq l \leq L-1 . \tag{4.7}$$

Usually codes are compared using the periodic auto- and cross-correlation functions, because they can be better investigated analytically than the aperiodic ones. But due to the random data bits to be transmitted, in the receiver of real communication systems in general the aperiodic correlation function appears. The properties of the aperiodic correlation function are usually worse than that of the periodic one.

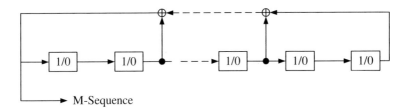

Fig. 4.1. Schematic of a code sequence generator

For a CDMA system, codes should have a high autocorrelation peak at $l = 0$. For all other values of l, $\phi_{xx}(l)$ should be as close to zero as possible. This aids in acquisition and tracking and gives reliable symbol decisions. For a high user capacity in the CDMA system the different codes must be mutually orthogonal, resulting in the condition for the cross-correlation function

$$\phi_{xy}(l = 0) = 0. \tag{4.8}$$

For any known code both auto- and cross-correlation properties cannot be fulfilled simultaneously [69]. Additionally the condition (4.8) is sufficient only if all signals are received time (data symbol) aligned. This can only be true for the downlink while in the uplink all user signals have different propagation paths to the base station and are no longer time aligned. In this case (4.8) should hold for all l. While $\phi_{xy}(l = 0) = 0$ is fulfilled for certain types of codes, low values for the cross-correlation function for $l \neq 0$ are not guaranteed in most cases.

4.1.2 M-Sequences

In most cases the spreading codes are generated with a shift register together with an appropriate logic, which feeds back a combination of the states of two or more of its stages to the input. Figure 4.1 shows a principal schematic. The logic realizes an exclusive or (*EXOR*) combination (a modulo 2 addition is equivalent to an EXOR operation in the case of binary codes) of the stages to be fed back. M-sequences are codes which have the maximum length (from which the name m-sequence or maximal-sequence arises) realizable with a shift register consisting of N stages, which is $L = 2^N - 1$. To achieve an m-sequence only certain shift register stages can be fed back. They are described by so-called *irreducible* or *primitive polynomials* [52, 69]. For different initial register values the resulting m-sequence is always the same – it just starts with a different time offset or so-called *code phase*.

The properties of m-sequences are very well investigated (e.g., [70]). Here we will list only the most important ones.

- For the autocorrelation function we find

$$\phi_{xx}(l) = \begin{cases} L & \text{for} \quad l = 0 \\ -1 & \text{for} \quad 1 \leq l \leq L - 1 \end{cases} \tag{4.9}$$

if the code elements $\{0,1\}$ are mapped to $\{-1,1\}$. This shape of the autocorrelation function results in a noise like (constant) PSD. The periodic and aperiodic autocorrelation function of an m-sequence of length 127 is plotted in Fig. 4.2.

- For the aperiodic autocorrelation function the ratio of the main peak value to the maximum sidelobe is approximately $\sqrt{2^N - 1}$.
- There are 2^{N-1} ones and $2^{N-1} - 1$ zeros within one code period. This results in a DC offset which prevents perfect carrier suppression.
- It has been shown ([52]) that the number of *runs* (a certain number of consecutive chips with equal value) with length p is $2^{n-(p+2)}$ for both ones and zeros. There are no runs of zeros with length N or of ones with length $N - 1$. The number of runs, therefore, decrease with the power of two as their length increases. The statistical distribution of ones and zeros appears to be totally random but is well defined and always the same. This randomness can also be seen from the shape of the autocorrelation function. The correlation of an m-sequence with a time-shifted version of its own is almost zero, indicating that the code values $c(i)$ are statistically independent.
- If an m-sequence is modulo-2 added to a time-shifted version of itself, the result is the same code sequence with a new time-shift. If two different m-sequences of equal length are added, the result is a composite sequence of the same length. The composite sequence is different for each combination of time-shifts between the two original sequences. This results in the ability to produce a large number of different codes, which is exploited for the generation of the Gold code family (section 4.1.3).
- It has been shown [69, 70] that for all m-sequences of a certain length, only a few of them have mutually "good" cross-correlation functions. In this context "good" means that the cross-correlation function takes only three different values, namely $\{-1, -\Theta_c(N), \Theta_c(N) - 2\}$, where

$$\Theta_c(N) = 2^{\lfloor N/2+1 \rfloor} + 1. \tag{4.10}$$

Here $\lfloor x \rfloor$ denotes the integer part of the real number x. Unfortunately the number of m-sequences with these "good" cross-correlation properties, known as the *mightiness*, is very small and does not increase with the length of the m-sequences. This is shown in the following Table 4.1. An algorithm to find all "good" m-sequences of a certain length out of all possible ones of the same length is described in [69] and [70].

The small mightiness of the m-sequences prevents their use for differentiating between users in CDMA systems. But their good autocorrelation property makes them well suited for synchronization purposes in all types of SS systems.

Table 4.1. Properties of m-sequences for different lengths of the generating shift register

N	3	5	7	11	15
L	7	31	127	2047	32767
Number of m-sequences	2	6	18	167	1800
Mightiness	2	3	6	4	2

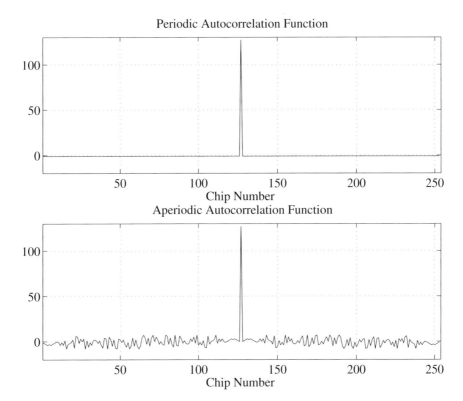

Fig. 4.2. Periodic and aperiodic autocorrelation function of an m-sequence of length 127

4.1.3 Gold Sequences

Gold sequences are generated by a chip-wise modulo 2 addition of two "good" (in the sense described in the preceding section) m-sequences of equal length. This is schematically shown in Fig. 4.3. Despite the low number of possible "good" m-sequences a high number of Gold sequences is available. This is, because the two m-sequences can have $2^N - 1$ different relative offsets between them. And each *different offset* results in a *different Gold sequence*, as we already stated when describing the properties of m-sequences.

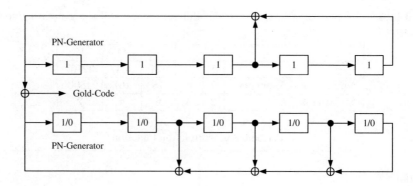

Fig. 4.3. Gold sequence generator

As opposed to m-sequences, the periodic autocorrelation function of Gold codes has four different values and the periodic cross-correlation function possesses three.

$$\phi_{xx}(l) \in \{L, \Theta_c(N) - 2, -1, -\Theta_c(N)\} \tag{4.11}$$

$$\phi_{xy}(l) \in \{\Theta_c(N) - 2, -1, -\Theta_c(N)\} \tag{4.12}$$

Here $\Theta_c(N)$ is the same as defined in (4.10). Comparing the Gold sequences to the m-sequences with respect to the correlation function, for the Gold sequences we find a much higher number of sequences with the same bounds on the cross-correlation function. This is compensated by a worse autocorrelation function whose sidelobes are as high as the maximum of the cross-correlation function. Gold sequences are used in UMTS for scrambling purposes.

4.1.4 Walsh Sequences

In 1923, J. L. Walsh defined a system of orthogonal functions [71]. Besides many other properties [57], the perfect orthogonality of the Walsh function is the most important feature for communications applications. Walsh sequences are used in UMTS as channelization codes in up- and downlink. We first describe the generation of Walsh sequences, and from this some properties of these sequences can easily be derived.

Walsh sequences can be generated with the help of so-called Hadamard matrices, which are square matrices. Other methods to generate Walsh sequences are described in [57]. Each row or column of a Hadamard matrix $\mathbf{H}_i$ is a Walsh sequence. The Hadamard matrices are computed using the following recursion:

$$\mathbf{H}_{2i} = \begin{bmatrix} \mathbf{H}_i & \mathbf{H}_i \\ \mathbf{H}_i & -\mathbf{H}_i \end{bmatrix}. \tag{4.13}$$

Starting point for the recursion is the Hadamard matrix of order 2

$$\mathbf{H}_2 = \begin{bmatrix} 1 & 1 \\ 1 & -1 \end{bmatrix}. \tag{4.14}$$

As an example we compute the Hadamard matrices of order 4 and 8:

$$\mathbf{H}_4 = \begin{bmatrix} 1 & 1 & 1 & 1 \\ 1 & -1 & 1 & -1 \\ 1 & 1 & -1 & -1 \\ 1 & -1 & -1 & 1 \end{bmatrix} \tag{4.15}$$

$$\mathbf{H}_8 = \begin{bmatrix} 1 & 1 & 1 & 1 & 1 & 1 & 1 & 1 \\ 1 & -1 & 1 & -1 & 1 & -1 & 1 & -1 \\ 1 & 1 & -1 & -1 & 1 & 1 & -1 & -1 \\ 1 & -1 & -1 & 1 & 1 & -1 & -1 & 1 \\ 1 & 1 & 1 & 1 & -1 & -1 & -1 & -1 \\ 1 & -1 & 1 & -1 & -1 & 1 & -1 & 1 \\ 1 & 1 & -1 & -1 & -1 & -1 & 1 & 1 \\ 1 & -1 & -1 & 1 & -1 & 1 & 1 & -1 \end{bmatrix}. \tag{4.16}$$

It can be easily seen that all rows (and all columns) are mutually orthogonal. The following properties of Walsh sequences can be derived if we define the Walsh sequence W_i as the ith row (or column) of a Hadamard matrix:

- Walsh sequences are binary sequences taking the values $+1$ and -1.
- The length of Walsh sequences is always a power of 2.
- There are always L different sequences of length L.
- Walsh sequences are mutually orthogonal if they are synchronized, i.e., $\phi_{xy}(l = 0) = 0$.
- If two Walsh sequences do not have zero time shift ($l \neq 0$) the periodic cross-correlation function can take values up to the autocorrelation peak, which is equal to the length L of the sequence. But it is also possible that the periodic cross-correlation function is zero for all time shifts. If we consider the above example of Walsh sequences of length 8, we find the following periodic cross-correlation sequences (not all possibilities are shown):

$$\phi_{W_2,W_{\{1,3,4,...8\}}} = [0 \quad 0 \quad 0 \quad 0 \quad 0 \quad 0 \quad 0 \quad 0] \tag{4.17}$$
$$\phi_{W_3,W_4} = [0 \quad -8 \quad 0 \quad 8 \quad 0 \quad -8 \quad 0 \quad 8] \tag{4.18}$$
$$\phi_{W_6,W_7} = [0 \quad 4 \quad 0 \quad -4 \quad 0 \quad -4 \quad 0 \quad 4] \tag{4.19}$$
$$\phi_{W_6,W_8} = [0 \quad 4 \quad -8 \quad 4 \quad 0 \quad -4 \quad 8 \quad -4]. \tag{4.20}$$

- The behavior of the aperiodic cross-correlation function is similar to that of the periodic one. Depending on which of the Walsh sequences are correlated the function can stay between ±1 but can also take values up to $\pm L - 1$.
- Every sequence starts with $+1$.

4.2 Receiver Structures

4.2.1 Correlation Receiver

In principle, the optimum receiver in CDMA systems is the same correlation or matched filter receiver as already described in section 3.2.2. Again, we have to correlate the received signal with all possible transmit signals, compare the results, and choose the signal which gives the highest correlation value. The optimum receiver was derived under the assumption that we were looking for a known signal corrupted by AWGN. In the case of a CDMA system we also have other interference. This interference results from the signals of other users which are active in the same frequency band and at the same time, because this is the characteristic of a CDMA system. Therefore, this is called *MAI*. To determine how MAI influences the performance of a CDMA system, we will consider a correlation receiver by recalling (3.13), which describes the correlation operation performed in the receiver,

$$\int_0^{T_D} r_L(t)\hat{s}_L(t)\mathrm{d}t = \int_0^{T_D} s_L(t)\hat{s}_L(t)\mathrm{d}t + \int_0^{T_D} \tilde{n}(t)\hat{s}_L(t)\mathrm{d}t\,. \tag{4.21}$$

Here, the received signal $r_L(t)$ is correlated with all possible transmit signals $\hat{s}_L(t)$. The interference signal $\tilde{n}(t)$, which is added to our transmit signal, is now composed of two terms: first, the AWGN signal $n(t)$, and second, $\tilde{s}_L(t)$, representing the MAI

$$\tilde{n}(t) = n(t) + \tilde{s}_L(t)\,. \tag{4.22}$$

The MAI $\tilde{s}_L(t)$ is the sum of signals from all other users in the same band which are active at the same time,

$$\tilde{s}_L(t) = \sum_{j=1}^{J-1}\sum_{n=-\infty}^{\infty} a_j d_j(n) \sum_{i=0}^{L-1} c_{j,i} \cdot g_C(t - iT_C - nT_D - T_j)\,. \tag{4.23}$$

Here, J is the number of all active users including the wanted user, $d_j(n)$ are the data symbols of the jth user, and a_j and T_j are the scaling and the time shift, respectively, of the jth user with respect to the wanted user for whom we assume the index $j = 0$. With the possible different scaling and time shift associated to each user, we account for the general case. This occurs if the user signals are not time-aligned and the users are located at different distances from the base station. In the UMTS downlink the time alignment is assured, while in the uplink the user signals are not aligned in the FDD mode. In the TDD mode, the uplink user signals are aligned within ± 4 chips or $\pm 1/4$ chip (see section 7.7.2). The situation is reversed for the scaling of the user signals. In the uplink we have power control (see section 4.3) to ensure nearly the same power for all received signals. Because the downlink user signals experience different mobile radio channels, they are transmitted with different power levels according to their channel.

In the multiuser case, to achieve the same receiver performance as for a single user, the correlation between the known transmit signals and the interference signal $\tilde{n}(t)$ must vanish. Considering (4.21), (4.22), and (4.23), we find that the correlation receiver computes the aperiodic cross-correlation functions between the code sequence of the wanted user and all other interference users. In the ideal case, i.e., if the codes are perfectly orthogonal, the interference user signals are time aligned and all signals have the same power, all cross-correlation terms in (4.23) vanish, and the receiver operates with the same performance as in the single user case. In real systems, however, the codes are usually not perfectly orthogonal and are neither time-aligned nor have all signals the same power. This justifies taking the auto- and cross-correlation functions as a performance measure for the code sequences. In addition, non-ideal characteristics of the components in the transmitter and receiver (e.g., nonlinearities, I/Q mismatch of quadrature modulators) cause additional deviations from orthogonality. Therefore, MAI places a fundamental limit on the performance of CDMA systems.

In a highly simplified manner MAI can be accounted for by the orthogonality factor OF, introduced in section 3.2.1 in (3.6), which we repeat here,

$$\frac{E_b}{I} = \frac{E_c}{I} + 10\log(G_{\mathrm{P}}) + OF . \tag{4.24}$$

E_b/I, E_c/I, and G_{P} are the bit energy to interference ratio, the chip energy to interference ratio, and the processing gain, respectively. If a noise-free environment is assumed, all codes are perfectly orthogonal, and despreading is performed ideally, OF tends toward infinity, stating that the despreading operation totally removes the MAI (which is the only source of interference under these assumptions). Without MAI and in the presence of AWGN, OF is zero, and in the general case (AWGN, MAI, and non-ideal despreading), OF becomes negative, which decreases the achievable E_b/I in the receiver.

We will now specify the performance of a correlation receiver in a CDMA system in terms of BEP under the following assumptions: the CDMA signals are BPSK modulated, the receiver is perfectly synchronized to the wanted user signal, the $J-1$ interfering user signals all arrive at the same power level as the wanted signal (perfect power control), the T_j's are uniformly distributed in the interval $[0, T_{\mathrm{D}}]$, the data bits d_j are independent, identically distributed random variables with $\Pr(d_j = +1) = \Pr(d_j = -1) = 0.5$, and the transmission channel just adds AWGN (no multipath propagation). For this scenario the sampled output of the correlation receiver is modeled according to [36]. To proceed any further, knowledge of the actual used spreading sequences (i.e., their cross-correlation properties) and the pulse shaping filter is necessary. If, for simplicity, MAI is modeled as Gaussian distributed (for which the number of interference user should be large), the following expression can be derived [36, 65]:

$$P_b = Q\left(\left[\frac{J-1}{3L} + \frac{N_0}{2E_b}\right]^{-1/2}\right). \tag{4.25}$$

We recall L as the processing gain (the length of the spreading sequences), $N_0/2$ as the two-sided power spectral density, and E_b as the energy per bit (= symbol in the case of BPSK). Because this approximation can be quite inaccurate for a small number J of interfering users, one derives the so-called improved Gaussian approximation [36]

$$P_b = \frac{2}{3} \cdot Q\left(\left[\frac{J-1}{3L} + \frac{N_0}{2E_b}\right]^{-1/2}\right) + \tag{4.26}$$

$$+ \frac{1}{6} \cdot Q\left(\left[\frac{J-1}{3L} + \frac{\sqrt{3}M}{L^2} + \frac{N_0}{2E_b}\right]^{-1/2}\right) +$$

$$+ \frac{1}{6} \cdot Q\left(\left[\frac{J-1}{3L} - \frac{\sqrt{3}M}{L^2} + \frac{N_0}{2E_b}\right]^{-1/2}\right)$$

with

$$M^2 = (K-1)\left[L^2 \frac{23}{360} + L\left(\frac{1}{20} + \frac{K-2}{36}\right) - \frac{1}{20} - \frac{K-2}{36}\right]. \tag{4.27}$$

Equation (4.26) is plotted in Fig. 4.4 for a processing gain (length of spreading code sequence) of $L = 127$. At low values of E_b/N_0 the BEP is determined by the AWGN. For higher values the BEP flattens out and cannot be decreased by an improvement of E_b/N_0. The saturation level and the value of E_b/N_0 where the saturation begins depend on the number of active users. The more users, the higher the BEP becomes. Therefore, it can concluded that CDMA systems are interference limited and not noise limited.

4.2.2 RAKE Receiver

In the above section an AWGN channel was assumed for the performance evaluation of a correlation receiver in a CDMA system. This is definitely a much too optimistic approach in a mobile communications scenario. Owing to the mobile radio channel (see section 2.3), the received signal is not only impaired by AWGN and MAI, but also strongly affected by fading effects due to multipath propagation and the Doppler effect. In CDMA systems with a transmission bandwidth larger than the coherence bandwidth B_C (section 2.3.4) of the channel (which is the case in UMTS), the transmission signals experience frequency selective fading. In this case, the impulse response model (section 2.3.3) for the channel can be applied. Some form of equalization technique in the receiver should be used (see, e.g., [6, 19, 72]) to account for the frequency dependent channel transfer function, as is usually done in narrow-band systems. The same is possible in CDMA systems, but they can

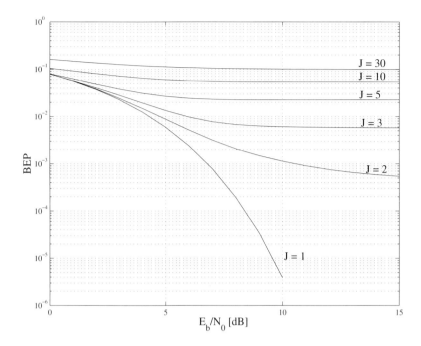

Fig. 4.4. Bit error probability for a CDMA system over an AWGN channel with an improved Gaussian approximation of multiple access interference as a function of E_b/N_0. The parameter of the plot is the number J of active users. A processing gain (length of spreading code sequence) of $L = 127$ was assumed

deal much better with the distortions resulting from the multipath effect. In fact, CDMA systems allow the exploitation of the multipath propagation to enhance the performance of the receiver. This becomes possible because of the autocorrelation properties of the spreading sequences. We assume a perfect or at least a "very good" autocorrelation function (i.e., $\phi_{xx}(0) = L$, $\phi_{xx}(\tau \neq 0) = \varepsilon$, and ε is very small compared to L). At the receiver's antenna we have the superposition of several replicas of the transmit signal, each with a different path delay and a different complex path coefficient. If the propagation times for the multiple paths differ by at least a chip duration from each other, a correlation receiver that is perfectly synchronized to one of the paths will suppress all other paths according to the autocorrelation property of the spreading sequence. This combats the fading due to the multipath effect, but leaves the problem of time variations of the path coefficient unresolved, e.g., if the path to which the receiver is synchronized exhibits sudden shadowing, the received signal could be lost.

A better approach to deal with multipath distortions is to use several correlation receivers in parallel with each of them synchronized to a different path. Since the mobile radio channel can feature a high number of paths and

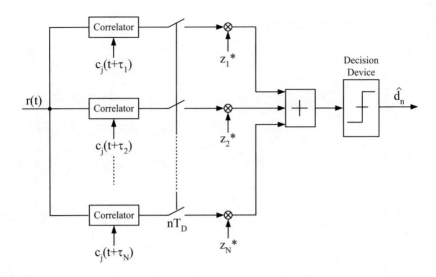

Fig. 4.5. A RAKE receiver with N fingers and maximum ratio combining

each correlation receiver needs a certain implementation effort, only a limited number of parallel receivers is implemented and they are synchronized to the strongest paths. The outputs of the multiple receivers are combined in a suitable manner to form a decision variable which is fed to the receiver's decision device. This type of receiver applies *multipath diversity*. Using this concept, the loss (or strong attenuation) of one path causes a much lower deterioration of the SNR of the decision variable as in the case of a receiver that uses just one path. Because of the triangular shape of the ideal autocorrelation function of the code sequence, only paths with propagation times that differ at least one chip duration from each other can be detected in different correlation receivers.

A receiver of this type is called a RAKE receiver because of its similarity to a garden rake. The principle block diagram of a RAKE receiver is shown in Fig. 4.5. The correlation receivers operating in parallel are called the RAKE fingers. Each of the finger is collecting the amount of signal energy received over the path to which this finger is synchronized. Therefore, the combined output of the fingers has a higher SNR as each of the fingers can achieve. The combining of the different RAKE finger outputs can be done using different algorithms, e.g., selective combining, switched combining, equal gain combining or maximum ratio combining [24]. Maximum ratio combining gives the highest SNR for the decision variable. It is applied if the correlator outputs are weighted with the complex conjugate path coefficients as shown in Fig. 4.5. If multiple antennas are applied at the receiver, their signals can be combined in the same way, i.e., by adding additional fingers to the RAKE receiver. In this case the fingers processing the signal of one antenna must track only the different paths of the mobile radio channel associated with this antenna.

For a performance estimation of a RAKE receiver we assume perfect knowledge (and tracking in the case of time variations) of the multipath components (path delay τ_i and path loss z_i), perfect synchronization of each RAKE finger, each path is subject to Rayleigh fading, the mean powers received over each path are all equal resulting in the same mean SNR_p for each path, and only one user is present in the system. The analysis of a RAKE receiver under these assumptions is the same as considering multipath diversity. This results in the following expression for the BEP as a function of the SNR of the whole channel (i.e., $SNR = N \cdot SNR_p$) and the number of RAKE fingers (or diversity branches) N [6, 65]:

$$P_b = \left(\frac{1-x}{2}\right)^N \sum_{i=0}^{N-1} \binom{N-1+i}{i} \left(\frac{1+x}{2}\right)^i \tag{4.28}$$

with

$$x = \sqrt{\frac{SNR_p}{1 + SNR_p}} . \tag{4.29}$$

In Fig. 4.6 the bit error performance of a RAKE receiver according to (4.28) is plotted. We find that changing from a conventional correlation receiver to a RAKE with 2 fingers results in the most significant performance gain of about 10 dB for a BEP of 10^{-3}. For each additional RAKE finger the performance improvement decreases. This usually limits the number of RAKE fingers in practical implementations to 3 to 5. A second limit for the number of RAKE fingers is the implementation complexity of a RAKE receiver, because a conventional correlation receiver including acquisition and tracking has to be realized for every finger. Today, RAKE receivers are used in almost every CDMA system. One of the RAKE fingers is used to constantly search for a path with a higher signal energy than the already processed path with the lowest energy.

4.2.3 Advanced Receiver Structures

Besides the RAKE receiver described above, more advanced receiver structures are possible for CDMA systems. The problems already addressed with conventional CDMA receivers are their lack of specialized treatment of MAI and the necessary power control. Since the interference from other users is known, or at least can be estimated in CDMA systems, it is possible to account for this interference in the receiver instead of considering it as uncorrelated noise as is done in a conventional RAKE receiver. This leads to the concept of *multiuser detection* [73]. While a detailed treatment of this topic, which has been a very popular area of research during the last decade, is beyond the scope of this book, we will give a very short overview of the

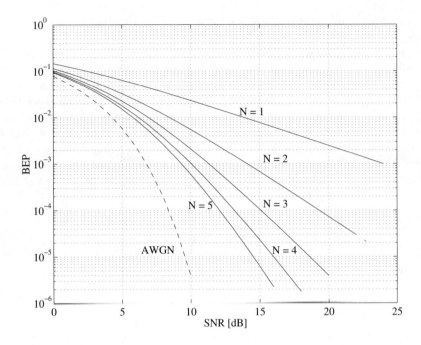

Fig. 4.6. Bit error performance of a RAKE receiver over a multipath channel with each path subject to Rayleigh fading. The parameter of the plot is the number of RAKE fingers N

different types of multiuser detection (MUD), since this meanwhile state-of-the-art technique will be applied in UMTS.

According to [74] the data detection techniques for DS-SS-based CDMA systems can be categorized as shown in Fig. 4.7. Single-user detection is defined as a method which demodulates the signal of a single user without requiring information on other interfering users. The conventional single-user detector uses the RAKE structure. There exist some advanced single-user structures that achieve a performance gain compared to the conventional RAKE receiver, which have similar computational complexity [74]. The multiuser detectors, which make use of knowledge or estimates of the signals of the other (interfering) users, are divided into *interference cancellation* (IC) and *joint detection* (JD) techniques. In addition some combined algorithms are possible but will not be outlined here.

Joint Detection. The structure of joint detection receivers is usually (but not necessarily) a bank of correlators followed by a linear or nonlinear transformations. The computational load is often high because of matrix calculations and inversions. The *Maximum Likelihood Sequence Estimator* shows optimum performance but suffers from high complexity with the number of operations growing in the order of 2^K with K as the number of users. It

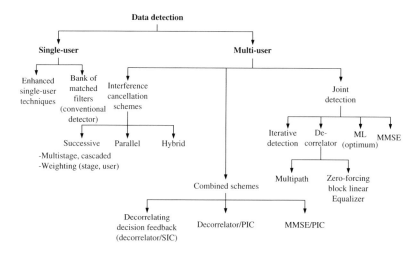

Fig. 4.7. Classification of data detection techniques for CDMA systems

consists of a Viterbi decoder placed after a bank of matched filters (each of them matched to one user).

The complexity of the *Decorrelator Detector* increases linearly with the number of users. Two main versions of this detector exist. The *Multipath Decorrelator Detector* views each received signal (the number of signals is K users times N paths) as independent interference source and calculates their correlation matrix. The inverse of the correlation matrix is multiplied with the output of the correlator bank, which removes the interference. A drawback is the noise enhancement that can be considerable. For the *Zero-Forcing Block Linear Equalizer* (ZF-BLE) interference is the result of the convolution of the interfering users transmit signals with their channel impulse responses. The receiver computes the inverse of the matrix of the convolution between the transmitted signals and the channel impulse responses. While more complex than the multipath decorrelator detector, the ZF-BLE has superior performance, but is highly dependent on the quality of the channel estimates.

The *Minimum Mean Square Error (MMSE) Detector* calculates the inverse of correlation matrix but according to the MMSE criterion takes the noise into account, too. With the same complexity, this gives a better BER performance than the multipath decorrelator detector if the system noise is not negligible and the received powers are known.

The *Iterative Multiuser Detector* applies a detection strategy based on that used for turbo decoding. The output of the bank of matched filters is used as input for an iteration process which, finally, gives the estimated data symbol. The algorithm is highly complex, since each iteration requires the manipulation of large matrices.

Interference Cancellation. IC requires the estimation and regeneration of the interference to finally be able to cancel the interference from the received signal. This needs knowledge (or estimation) of the channel parameters and the interfering data. Errors in both estimates degrade the performance or can even increase the interference after the IC.

In the *Successive Interference Cancellation* (SIC) scheme first all users are detected using a bank of correlators. Then the users are ranked according to their received signal strength. Only the data from the strongest user are used to regenerate its associated (interfering) signal which is fed back and subtracted from the received signal. This process is repeated until all interfering users have been detected and canceled. While the successive removal of just one user in each step needs only low computational effort the associated delay can be quite long. Further issues are different BER's for different users due to unequal signal-to-noise-and-interference ratios and error propagation if interfering users are incompletely or incorrectly canceled.

The alternative approach is *Parallel Interference Cancellation* (PIC). Using this scheme the interference from all users K to the other $K - 1$ users is estimated and canceled simultaneously. After the IC, the data symbols are detected. Since these data are more reliable than before the IC, they can be used to generate a better estimate of the interference leading to a second PIC stage. With the increasing number of stages the data decisions become better but the computational complexity increases, too. Compared to the SIC, the delay due to signal processing is low. The PIC can completely cancel the interference if the first interference estimations are correct. A modification of the PIC is the use of soft decisions instead of hard decisions

The idea to combine the advantages of SIC and PIC leads to the *Hybrid Interference Cancellation* (HIC). One possibility is to divide the K users into groups (preferably groups of user signals with nearly equal strength). In each group PIC is performed and the groups are processed serially. Another approach is to cancel the interference partially in parallel and partially in series, e.g., in each step one PIC stage is performed on the P strongest users and afterwards the S strongest users are canceled serially. This scheme is repeated until sufficient reliability is reached.

4.3 Power Control

Although power control in an issue not confined to CDMA systems (e.g., there is a power control scheme in GSM) we will review this topic briefly with respect to CDMA. In general, power control means controlling the transmit power to optimize the receiver performance for the case that many users are active in the same cell and in the neighboring cells. This leads to an optimization of the system/cell capacity.

We first consider the *uplink scenario*. If all mobile stations would transmit with the same power (no power control) the received signal of a MS near the base station would be much stronger than a signal from a MS far away. In the case of perfect cross-correlation properties of the spreading sequences an ideal correlation receiver could recover the weak signal, but the receiver does not operate ideally nor do the sequences have perfect properties. Thus, a CDMA system is interference limited (section 4.2.1) and the interfering signal of the nearby MS (MAI) could completely block the reception of the weak signal. This is called the *near–far problem*. Obviously, the solution to this is some form of power control. With perfect power control, the transmit powers of all MSs in a cell are controlled in way that their transmit signals are received at the BS's antenna with exactly the same power. Of course, in this case we still have MAI because each signal causes interference with respect to all other users. But now each user adds only a small fraction of MAI to the other users signals. Therefore, power control can be viewed as distributing the overall acceptable MAI allowed in the system to the highest possible number of simultaneous users. Without power control one MS could use up all allowed interference, thereby blocking other users from being served by the system.

The situation is different in the *downlink*. Because of the different path losses for each MS, the BS transmits the different user signals with different power levels. With perfect power control the signals arrive at their associated MS with exactly the same power level. This would *create a near–far problem* because the transmit signal for a MS near the BS would be much weaker than that of a MS at the cell boundary and, therefore, could be blocked. Nevertheless, in UMTS, downlink power control is applied, but the dynamic rage is much smaller than in the uplink and the goal is different. Downlink power control is used to provide a power margin for links to MSs at the cell edges where they suffer from inter-cell interference. Some additional transmit power can be provided also for mobiles at low speeds, which are in deep fades and have established high data rate links [75].

In the following sections we distinguish between open loop, closed loop, and outer loop power control mechanisms.

4.3.1 Open Loop Power Control

With open loop power control the transceiver attempts to maintain the sum of received and transmitted power at a constant level. If the received power decreases, the transmit power is increased and vice versa. Since up- and downlink transmission take place in different frequency bands in FDD systems, up- and downlink path loss are not usually the same and the fast fading is definitely uncorrelated in both bands. Therefore, this type of power control has only poor precision. It is used to provide a coarse initial setting of the transmit power.

4.3.2 Closed Loop Power Control

In a closed loop power control the BS controls the MS's transmit power via a control channel. In the base station the received signal-to-interference ratio SIR is measured and compared to a target SIR. Depending on the result an appropriate power control command is issued to the MS. The speed of the power control (the frequency at which the MS receives a power control command) must be faster than the Rayleigh fading to compensate for these changes. The compensation of channel fading by controlling the MS's transmit power is advantageous for the receiver in the BS but results in an increased average transmit power and introduces enhanced interference to neighboring cells. Thus, a compromise has to be found.

Often the MS's power is changed by a fixed step size with every power control command. In this case there must be a tradeoff between the speed of the power control (which requires a large step size) and stability and small error (which both require a small step size). In most cases both open and closed loop power control schemes are used together in a system.

4.3.3 Outer Loop Power Control

Outer loop power control is closely connected to the closed loop power control. Outer loop power control adjusts the target SIR of the closed loop power control to maintain a constant link quality, usually defined in terms of bit error rate (BER) or frame error rate (FER). This adjustment is necessary, since the required SIR for a certain FER depends on parameters like the multipath profile and mobile speed. It would not make sense to specify a target SIR for the worst case, because this would result in a transmit power which is too high under good channel conditions.

5. Introduction to UMTS

In this chapter we give an introduction into the background of 3G mobile communication systems. We will start with a general definition and the basic concept. A short section is devoted to market aspects to acquaint the reader with the economic facts and expectations behind 3G. A description of spectrum issues follows before we highlight the currently ongoing standardization process and the organizations involved.

5.1 Definition and Concept of Third Generation Mobile Communication Systems

In 1985 the ITU started work on 3G systems under the acronym FPLMTS (Future Public Land Mobile Telephone System) which was later renamed IMT-2000 [76]. The key factors and main objectives for 3G systems can be summarized as follows:

- High-speed access (not only limited to wireless access) capable of bearing broadband services like Internet, video, and other multimedia services. Wireless services should support a minimum data rate of 144 kbps in all radio environments and up to 2 Mbps in low-mobility and indoor environments.
- Support of symmetrical and asymmetrical traffic.
- Packet access. Most of the traffic in 3G networks will originate from data communications. Therefore, packet switched communications must be provided in addition to a circuit switched mode to ensure efficient resource usage. This feature is introduced already with 2.5G systems like GPRS [77].
- Voice quality comparable to wire-line quality.
- High capacity and spectrum efficiency.
- Integration of residential, office, and cellular services into a single system based on one piece of user equipment.
- High flexibility to support new kinds of services. 3G systems will provide basic communications services, with applications and services being independent from the underlying transport layers.
- Multimedia service capability.

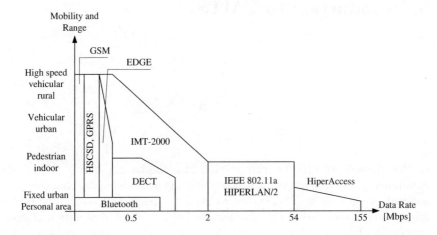

Fig. 5.1. Comparison of different 2G, 2.5G, 3G, and other wireless systems

- Low cost, at least for basic services.
- World-wide coverage and roaming. Even if a single world-wide standard is not feasible, it should become possible for the user to access the communications environment he wants from anywhere, no matter what type of terminal he uses. World-wide coverage also includes a satellite component.
- Evolution. The transition from 2G to 3G will be an evolutionary path. In the beginning, 3G systems and services must coexist with today's 2G and 2.5G systems, since nobody (users or network operators) would be able or willing to afford a hard transition from 2G or 2.5G to 3G.
- Capacity and capability to serve more than 50% of the population [78].

The general vision of 3G systems is that a user should be able to seamlessly roam amongst the various networks and radio environments with intelligent multi-mode handsets that are able to automatically detect the network and the radio interface they are experiencing and transparently select the appropriate mode. A world-wide coverage is envisioned by a network consisting of pico-cells for indoor and hot-spot applications, micro- and macro-cells, and world-wide satellite services. IMT-2000 is intended as a flexible standard for wireless access to the global telecommunications infrastructure which will serve both mobile and fixed users in both public and private networks. It should ensure communications from anywhere to anybody at any time.

From the above-listed items the basic demand for data throughput over the air interface was identified as 144 kbps (preferably 384 kbps) with full coverage and high mobility of the MS and up to 2 Mbps for low mobility and coverage limited to high traffic areas. These bit rates were harmonized to the ISDN (Integrated Services Digital Network) 2B+D (144 kbps), H0 (384 kbps), and H12 (1920 kbps) channels [76]. Figure 5.1 shows the data rates with their

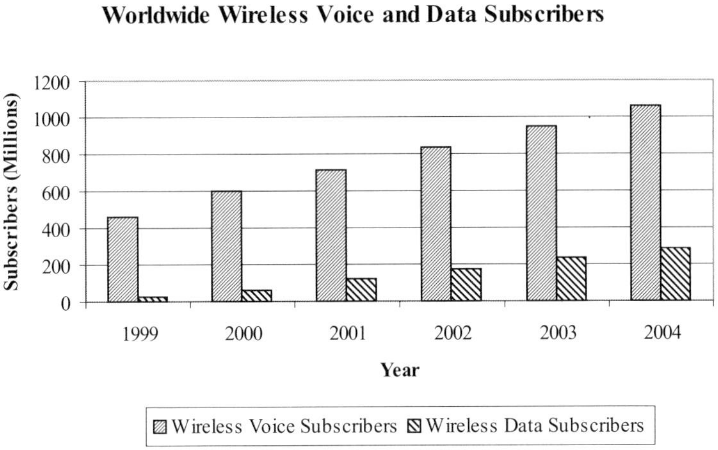

Worldwide Wireless Voice and Data Subscribers

Fig. 5.2. Number of mobile subscribers world-wide (*Source*: UMTS Forum)

maximum possible mobility for different 2G, 2.5G, and 3G systems, systems like DECT (Digital Enhanced Cordless Telecommunications) [79], Bluetooth [80] and WLAN-type systems such as IEEE 802.11 [81] and HIPERLAN/2 [42]. The evolvements of 2G systems like GPRS or EDGE, which are frequently termed 2.5G systems, as well as IMT-2000 will bring a distinct data rate improvement compared to 2G. For data rates above 2 Mbps specialized WLAN systems are foreseen.

5.2 Market Aspects

The first generation analog mobile communication systems (e.g., AMPS or NMT) were deployed starting in the early 1980s. Because these systems were limited to specific countries or regions, the number of subscribers as well as their economic success was limited. Second generation systems (e.g., the European GSM, IS-136 and IS-95, both starting in the US, or the Japanese PDC (Personal Digital Cellular)) are using digital technology, thus, providing additional services to the basic voice service. Besides these features most 2G systems provide a larger geographic coverage than first generation systems, moving one step further to global roaming. These systems were and still are a big economic success. The number of subscribers increased much faster than expected. Figure 5.2 shows the growth of world-wide mobile subscribers since 1999 with a forecast up to the year 2004. The number of mobile subscribers world-wide at the end of 1999 was about 470 million. For the year 2010 more than 1700 million mobile subscribers are expected [1]. Figure 5.3 shows

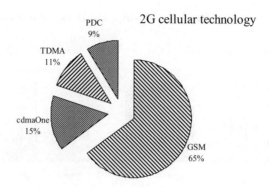

Fig. 5.3. 2G technology statistics estimated as of end of year 2000

the market share estimates of the 2G technologies for the end of the year 2000 [82]. GSM is by far the most accepted technology, capturing almost two thirds of the total 2G market. Despite their maturity and acceptance, 2G standards are still under continual development in order to incorporate new features and/or alleviate the major drawback of 2G systems, which is their low maximum data rate. Data traffic is limited to between 8 and 19.2 kbps, depending on the standard. For today's communications needs this is much too low, since data traffic is rapidly increasing. This is due to the strongly rising demand for wireless access to email and Internet services. It is expected that the market share of data services exceeds that of voice services before 2010. In addition, revenue from voice services will saturate while data services are expected to grow exponentially for the next several years as we are still at the beginning of this development. The data traffic originates from three, partly overlapping areas [83]:

- Telecommunications: This comprises voice and mobile ISDN services, video telephony and conferencing, and wideband data services.
- Computer-data: This includes email, Internet access, multimedia document transfer and mobile computing.
- Audio-Video: Under this item we summarize electronic newspaper, video on demand, interactive video, and TV and radio distribution.

Common to most of the listed applications is a strong asymmetry in the data traffic. Only a few bytes are sent from the MS to request a Web page or a video, while several megabytes are transmitted to the MS.

One of the major enhancements of GSM is the introduction of packet oriented connections with GPRS, which is just being deployed. GPRS should account for the increasing demand for data applications like email or Web browsing. Other developments for 2G systems are EDGE for GSM and NADC

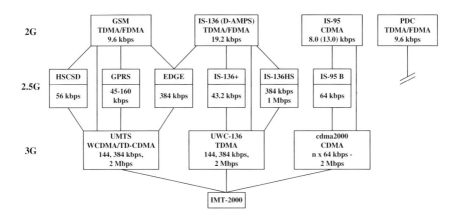

Fig. 5.4. Evolution from 2G systems towards 3G systems

(North American Digital Cellular) or IS-95B for IS-95 (see also Fig. 5.4). All these evolved 2G systems are frequently termed 2.5G systems. A review of these 2.5G systems can be found in [84] and, together with their interrelations to 2G and 3G systems, they are displayed in Fig. 5.4. Only for the PDC system there is no enhancement. In Japan the introduction of 3G systems will be a revolutionary step, while everywhere else an evolution from 2G to 3G should be possible. If the 2.5G systems are fully deployed, they will allow for data rates up to 384 kbps. With this infrastructure available, the mobile units will evolve from mobile phones to so-called smart phones including, e.g., PDAs (Personal Digital Assistants), to better support data applications like email, Internet or location based services. Also mobile Web panels are under development. If the maximum data rate of 2 Mbps in UMTS becomes accessible, such demanding applications like video conferencing could be supported. This will again introduce a new generation of mobile terminals capable of dealing with multimedia contents.

5.3 Spectrum Issues

The demand for high data rate services for many subscribers requires a large bandwidth to support these services. Figure 5.5 shows the expected spectrum requirements for the years 2005 and 2010 according to [85]. The above described rapid growth of data traffic with its asymmetry can clearly be seen.

In 1992, the WARC identified 230 MHz of spectrum for use with 3G mobile radio systems. In Fig. 5.6 this spectrum is shown together with the allocated IMT-2000 bands in Europe, China, Japan, Korea, and North America. Europe, Japan, China, and Korea basically implemented the WARC recommendation. Only in the lower part of the spectrum are there minor conflicts with DECT in Europe and PHS (Personal Handyphone System) in Japan.

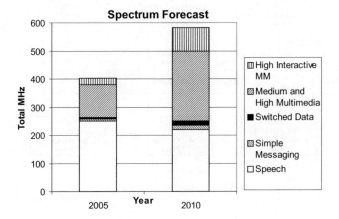

Fig. 5.5. Spectrum requirement forecast for terrestrial mobile communication services

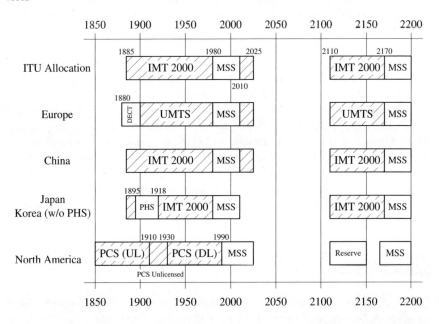

Fig. 5.6. IMT-2000 frequency allocation and their implementation in different regions of the world

In China, some of the lower part of the IMT-2002 spectrum is reserved for WLL (Wireless Local Loop) applications, but this has yet to be allocated to an operator. The situation is different in the USA. Here, most of the lower part of the IMT-2000 spectrum was assigned to second generation PCS (Personal Communication Systems) and has already been auctioned to operators. Thus 3G systems must be implemented within the existing bands. The 3G standardization tries to take care of that, by developing appropriate standard specifications (see the following section).

Due to the projected demand for spectrum in excess of the already allocated 230 MHz, the World Radiocommunication Conference 2000 (WRC-2000) identified additional bandwidth for the use with IMT-2000. These bands are:

- 1710 − 1885 MHz for terrestrial use,
- 2500 − 2690 MHz for terrestrial use,
- 806 − 960 MHz for terrestrial use and
- 1525−1544 MHz, 1545 − 1559 MHz, 1610 − 1626.5 MHz, 1626.5 − 1645.5 MHz, 1646.5 − 1660.5 MHz and 2483.5 − 2500 MHz for satellite services

The WRC-2000 further recognized, that so-called high altitude platform stations (HAPS) may offer a new means of providing IMT-2000 services with minimal network infrastructure as they are capable of providing service to a large footprint together with a dense coverage. HAPS are defined as stations located on an object at an altitude of 20 to 50 km and at a specified, nominal, fixed point relative to the Earth. In the resolution [COM5/13] the WRC-2000 defined some regulations to allow for the co-existence of HAPS with conventional terrestrial base stations and mobile IMT-2000 satellite stations [86].

5.4 Standardization

5.4.1 Development of 3G Systems

The work on IMT-2000 started in the ITU in 1985 under the term FPLMTS. In TG (Task Group) 8/1 of ITU-R (the Radiocommunications Sector) the radio aspects are considered, while the Telecommunications Standardization Sector (ITU-T) works on the radio-independent issues. Upon a request from the ITU, different radio transmission technology (RTT) proposals for IMT-2000 were submitted in 1998. The following regional standardization bodies were involved: the SMG (Special Mobile Group) from ETSI in Europe, the Chinese RITT (Research Institute of Telecommunications Transmission), the ARIB (Association of Radio Industry and Business) and TTC (Telecommunication Technology Committee) from Japan, the TTA (Telecommunications Technologies Association) from Korea, and from the USA, the TIA (Telecommunications Industry Association) and T1P1 [87]. Further details on structure and organization of these bodies can be found in [76]. In Table 5.1, the RTT proposals are listed. Details are available at [88]. It should be noted that eight of the 10 proposals are based on W-CDMA or at least contain a W-CDMA component.

Table 5.1. RTT proposals submitted to ITU

Proposal	Description	Source
CDMA I	Multiband synchronous DS-CDMA	S. Korea TTA
CDMA II	Asynchronous DS-CDMA	S. Korea TTA
cdma2000	Wideband CDMA (IS-95)	USA TIA TR45.5
DECT	Digital Enhanced Cordless Telecommunications	ETSI Project (EP) DECT
NA: W-CDMA	North American: Wideband CDMA	USA TIA T1P1-ATIS
TD-SCDMA	Time Division Synchronous CDMA	China Academy of Telecommunication Technology (CATT)
UTRA	UMTS Terrestrial Radio Access	ETSI SMG2
UWC-136	Universal Wireless Communications	USA TIA TR45.3
W-CDMA	Wideband CDMA	Japan ARIB
WIMS W-CDMA	Wireless Multimedia and Messaging Service Wideband CDMA	USA TIA TR46.1

During the development of ETSI's UTRA proposal, which was later adopted as one of the IMT-2000 family of standards, three key decisions were made [89].

- In January 1998 a consensus was reached in ETSI to integrate W-CDMA and time-division CDMA (TD-CDMA) as two modes (FDD and TDD) into one common UMTS standard. *Integrated* means that there are different physical layers for FDD and TDD, but layers 2 and 3 are common to both. This should ensure optimum service for all environments from high mobility in macro cells to low mobility in pico-cells in out- and indoor scenarios.
- The UMTS Terrestrial Radio Access Network (UTRAN) is built on ATM (Asynchronous Transfer Mode) AAL2 (ATM Adaptation Layer 2). ATM uses statistical multiplexing which is advantageous due to the variable bit rate traffic which results from the time varying radio interface. The use of AAL2 allows the guarantee of certain QoS parameters even for very low data rates or during soft handover, typical of CDMA-based systems.
- The existing GSM network is reused and enhanced to serve as part of the UMTS core network. The dominance of GSM among the 2G technologies (see Fig. 5.3) together with the annual releases of GSM 2+ standards which are developing towards 3G features, makes the GSM core network an ideal complement to the UTRAN, not only from the market perspec-

tive. In the initial phase of 3G there will be UMTS "islands" in a "sea" of GSM networks, which makes UMTS/GSM dual mode terminals necessary. Therefore, a combined GSM/UMTS network is a prerequisite for seamless service delivery.

The ARIB and ETSI proposals were almost identical, since both organizations worked closely together to develop a common standard. Because IMT-2000 was scheduled to start in Japan in 2001, ARIB froze the specification in the first half of 1999. The North American NA:W-CDMA and WIMS proposals were merged in a further step into WP-CDMA, a wideband packet W-CDMA. The W-CDMA proposals can be divided into two groups. One group with similar parameters includes W-CDMA, NA:W-CDMA (later WP-CDMA), TD-SCDMA (similar to the UTRA TDD), and CDMA II. Within this group, backwards compatibility to the existing GSM and PCS in the design of the protocols have been taken into account. The other CDMA-related group consists of CDMA I and cdma2000, which should provide the evolutionary path from the existing IS-95 system. The UWC-136 proposal from the USA represents an evolution from the 2G IS-136 system, while the DECT proposal relies on the current DECT system that has been mostly used in indoor applications up to now.

Besides the proposals for the terrestrial component of IMT-2000 there were also five proposals submitted for satellite RTT as can be seen from Table 5.2. The proposals from Inmarsat and ICO are based on TDMA as a

Table 5.2. Satellite RTT proposals submitted to ITU

Proposal	Description	Source
Horizons	Horizons satellite system	Inmarsat
ICO RTT	10 MEO sats in 2 planes at 10390 km	ICO Global Communications
SAT-CDMA	49 LEO sats in 7 planes at 2000 km	S. Korea TTA
SW-CDMA	Satellite Wideband CDMA	ESA
SW-CTDMA	Satellite Wideband hybrid CDMA/TDMA	ESA

multiple access scheme while the proposal from Korea and the SW-CDMA proposal from ESA (European Space Agency) rely on wideband CDMA. The second submission form ESA (SW-CTDMA) is a hybrid TDMA/W-CDMA technology. The W-CDMA-based RTT proposals are closely related to their terrestrial counterparts which is the UTRA FDD mode for SW-CDMA and CDMA II for SAT-CDMA. They represent an adaptation of the terrestrial RTT to the satellite environment. This close relation should provide a seamless integration of the satellite component of IMT-2000 into the whole 3G

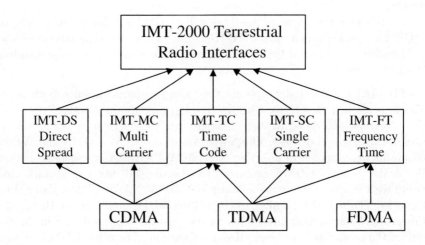

Fig. 5.7. The set of IMT-2000 terrestrial radio interfaces

system as envisaged by the ITU. Some basic features of SW-CDMA and SAT-CDMA are [78]:

- Information bit rates up to 144 kbps (SW-CDMA) or 128 kbps (SAT-CDMA).
- Same slot and frame structure as in UTRA FDD mode (see section 6.3).
- Essentially the same coding scheme, chip rate, modulation and pulse shaping filter as in UTRA FDD mode.
- Spreading with OVSF codes (see section 6.4.1) and possibility of multi-code transmission.
- Open loop power control for the random access channel and closed loop power control based on signal-to-noise and interference ratio for the traffic channels.

5.4.2 IMT-2000 Family of Standards

In a second phase the different terrestrial RTT proposals submitted to the ITU were evaluated. It turned out that the vision of a global standard with a single radio interface was not realizable for 3G systems. This was due to the different 2G technologies used in the different regions in the world. It would have been impossible to find one technology as an evolutionary path for all existing 2G systems. Therefore, a family concept was adopted and agreed upon at the end of 1999 [90]. The five standards included in IMT-2000 are shown in Fig. 5.7. As IMT-DS (Direct Spread) the UTRA FDD mode was adopted. IMT-TC (Time-Code) is a combination of the UTRA TDD and the TD-SCDMA proposals, cdma2000 is found in IMT-MC (Multi Carrier), IMT-SC (Single Carrier) corresponds to UWC-136, and IMT-FT (Frequency

Time) is the DECT proposal. These five standards are now further developed in the regional standardization bodies.

For the W-CDMA-based technologies (IMT-DS and IMT-TC) the 3rd Generation Partnership Project (3GPP) was created [91]. The following organizations are involved in 3GPP: ARIB, ETSI, TTA, TTC, T1P1, CWTS (China Wireless Telecommunication Standard Group). In addition, many equipment and semiconductor manufacturers and operators as well as market organizations (GSM Association, UMTS Forum, Global Mobile Suppliers Association, IPv6 Forum, Universal Wireless Communications Consortium (UWCC)) also joined 3GPP. A similar group was founded for the development of the cdma2000-based systems, termed 3GPP2 [92]. This activity is running in parallel to 3GPP and they are coordinated [75].

5.4.3 Sources

Table 5.3 lists the associated WWW addresses for each of the IMT-2000 standards.

Table 5.3. WWW addresses for the IMT-2000 standards

Standard	WWW	Comment
IMT-DS (UTR FDD)	www.3gpp.org	free access
IMT-TC (UTRA TDD)	www.3gpp.org	free access
IMT-MC (cdma2000)	www.3gpp2.org	free access
IMT-SC (UWC-136)	ftp.tiaonline.org/uwc136/	free access
IMT-FT (DECT)	www.etsi.org	free access

An overview of the documents used for the description of the UTRA FDD and TDD physical layer in chapters 6 and 7 is given in Table 5.4. All UMTS related details throughout this book are based on the June 2001, release 4 version of the technical specifications listed in the references. UMTS release 99, which was finalized at the end of 1999 is the basis for the first introduction of UMTS. According to [93], release 4, which is still under development, will add some functionality (e.g., the 1.28 Mcps option of the TDD mode, see chapter 7), but will involve no substantial changes in the core network. UMTS release 5, however, will migrate the core network to a single IP (Internet Protocol) network.

Table 5.4. 3GPP documents concerning the physical layer of UTRA FDD and TDD mode

Document	Title
3GPP TS 22.101	Service aspects; service principles
3GPP TS 22.105	Service aspects; services and service capabilities
3GPP TS 23.002	Services and systems aspects; network architecture
3GPP TS 25.101	UE radio transmission and reception (FDD)
3GPP TS 25.102	UE radio transmission and reception (TDD)
3GPP TS 25.104	BTS radio transmission and reception (FDD)
3GPP TS 25.105	BTS radio transmission and reception (TDD)
3GPP TS 25.141	Base station conformance testing (FDD)
3GPP TS 25.142	Base station conformance testing (TDD)
3GPP TS 25.201	Physical layer – general description
3GPP TS 25.211	Physical channels and mapping of transport channels onto physical channels (FDD)
3GPP TS 25.212	Multiplexing and channel coding (FDD)
3GPP TS 25.213	Spreading and modulation (FDD)
3GPP TS 25.214	Physical layer procedures (FDD)
3GPP TS 25.215	Physical layer – measurements (FDD)
3GPP TS 25.221	Physical channels and mapping of transport channels onto physical channels (TDD)
3GPP TS 25.222	Multiplexing and channel coding (TDD)
3GPP TS 25.223	Spreading and modulation (TDD)
3GPP TS 25.224	Physical layer procedures (TDD)
3GPP TS 25.225	Physical layer – measurements (TDD)
3GPP TS 34.121	Terminal conformance specification; radio transmission and reception (FDD)
3GPP TS 34.122	Terminal conformance specification; radio transmission and reception (TDD)

6. UTRA FDD Mode

This chapter provides a detailed description of the UTRA FDD mode. In the first section the key parameters are listed as a reference. A short section on the system architecture, the layer structure of the radio interface and its interworking with higher layers, and the bearer services provided in UMTS follows. The remainder of the chapter deals with the details of the physical layer, namely, transport and physical channels, spreading and modulation, coding and interleaving, physical layer procedures, compressed mode, and RF-characteristics. We will mostly use the terms MS or terminal instead of the 3GPP convention User Equipment (UE), and BS instead of Node B.

6.1 Key Parameters

The following Table 6.1 lists the key parameters of the UMTS FDD mode. A detailed description will be given in the remainder of this chapter.

6.2 System Architecture

The system architecture of UMTS is in principle the same as in 2G systems. Each logical network element has a well-defined functionality. The elements are connected via open interfaces, i.e., the interfaces are described in such detail that network elements from different manufacturers can be used together. In UMTS, the Radio Access Network (RAN) element is named UTRAN (UMTS Terrestrial RAN). The UTRAN handles all radio-related functionality. As a second network element we have the Core Network (CN), performing routing and switching of connections to external networks. The third network element is the UE. A picture of this high-level system architecture is shown in Fig. 6.1. Between UE and the UTRAN we find the Uu interface (which is the W-CDMA radio interface). The Iu interface connects UTRAN and the CN. Details concerning the interfaces can be found in the TS 25.41X documents for the Iu interface. Due to the W-CDMA technology used on the radio interface, the specifications for UE and UTRAN had (and still have) to be standardized completely new. The CN, however, is mainly

Table 6.1. Key parameters of the UMTS FDD mode

Parameter	Value	Description
B_{UL}	1920–1980 MHz	UL (uplink) frequency band
$B_{UL}{}^1$	1850–1910 MHz	UL frequency band in region 2
B_{DL}	2110–2170 MHz	DL (downlink) frequency band
$B_{DL}{}^1$	1930–1990 MHz	DL frequency band in region 2
Δf	5 MHz	Channel spacing
f_{chip}	3.84 Mcps	Chip rate
T_{chip}	0.26042 µs	Duration of a chip
N_{chip}	2560	Number of chips per slot
T_{slot}	0.666 ms	Duration of a slot
T_{frame}	10 ms	Duration of a frame
$T_{superframe}$	720 ms	Duration of a superframe
$f_{bit,inf}$	12.2–2048 kbit/s	Information bit rate
$N_{bit}(UL)$	10–640	Number of bits per slot (uplink)
$N_{bit}(DL)$	10–1280	Number of bits per slot (downlink)
$f_{bit,ph}(UL)$	15–960 kbit/s	Physical bit rate (uplink)
$f_{bit,ph}(DL)$	15–1920 kbit/s	Physical bit rate (downlink)
SF(UL)	256–4	Spreading factor (uplink)
SF(DL)	512–4	Spreading factor (downlink)
α	0.22	Roll-off factor (root-raised cosine)

[1] ITU Region 2 comprises the countries of North and South America.

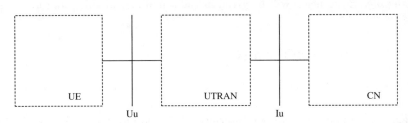

Fig. 6.1. UMTS high-level architecture

based on the GSM network of circuit-switched services and the GPRS overlay
for packet services [98, 99]. This will accelerate the introduction of UMTS
and enable roaming between the two systems. With release 5 of the UMTS
specification the CN will migrate to an IP (Internet Protocol) based network.
A more detailed description of the logical network elements is provided in,
e.g., [75].

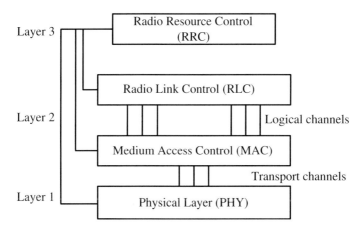

Fig. 6.2. Layer structure for the UTRA radio interface

6.2.1 Layer Structure

The UTRA radio interface comprises the three layers shown in Fig. 6.2. This book covers only layer 1, the physical layer (PHY), in detail. Functions and procedures associated with layer 1 are as follows: multiple access scheme, channel de/coding and de/interleaving, rate matching, frequency and time synchronization, modulation and spreading, power control, cell search, macro-diversity distribution/combining and soft handover execution, multiplexing of transport channels, and measurements of quantities like FER and SIR. Layer 2 is divided into the Medium Access Control (MAC) and the Radio Link Control (RLC) layers. Layer 3 is named Radio Resource Control (RRC). The functionality of layer 1 is offered to the MAC layer via the transport channels.

MAC Layer. In the MAC layer [94] logical channels which are offered to the RLC layer are mapped onto transport channels. Depending on the data rates necessary to serve the application chosen by the user a proper transport format is selected by the MAC layer. Further functions performed by the MAC layer are:

- Priority handling between MSs and between data flows of one MS.
- Multiplexing/demultiplexing of transport blocks to/from higher layers.
- Ciphering [95].
- Traffic volume monitoring. These measurements are used by the RRC layer to trigger reconfiguration of bearer services (see later).

RLC Layer. The function of the RLC layer [96] is to provide segmentation and retransmission services for user and control data. These services are called Signaling Radio Bearer (SRB) and Radio Bearer (RB). The RLC can operate in three different modes determined by the RRC. In the *transparent mode* no protocol overhead is added to the higher layer data and these are not segmented. If so-called protocol data units (PDUs) are erroneous, they are either discarded or marked. This is the same as in the *unacknowledged mode*, where no retransmission protocol is used and, therefore, data delivery is not guaranteed. The segmentation and concatenation of the data stream is performed by means of header fields added to the data. In the *acknowledged mode* an automatic repeat request (ARQ) protocol is used for error correction. The number of retransmissions is controlled by the RRC to determine the performance (a compromise between quality and delay). If the RLC cannot deliver its packets correctly despite the ARQ protocol, the upper layer is notified. Two types of packet delivery are possible. With in-sequence delivery the order of higher layer packets is maintained. Otherwise the higher layer PDUs are forwarded as soon as they are completely received with out-of-sequence delivery. The acknowledged mode is the usual mode for any packet-type data services such as data transfer or Internet browsing.

RRC Layer. The RRC layer [75, 97] performs most of the control signaling between UTRAN and UE. The RRC functions include:

- Broadcast of system information.
- Paging to set up a connection for a specific MS or change its service state. Different service states are possible for a MS if it is connected to the UTRAN, depending on the channels over which the MS can communicate with the UTRAN.
- Establishment, maintenance, and release of a RRC connection upon request from either MS or UTRAN.
- Establishment of Radio Bearers, transport channels and physical channels.
- Control of security functions (ciphering for data and signaling, and integrity protection only for signaling).
- Control of the measurements performed by the MS and reporting of the results to the UTRAN .
- Downlink outer loop power control (setting the SIR target).
- Open loop power control. The MS calculates the transmit power for the preamble before PRACH transmission (see section 6.3.2) depending on its own measurements and depending on system specific parameters received from the RRC.
- Mobility functions. A number of procedures are specified for the RRC to keep track of the cell in which a MS is located.

6.2.2 Services and QoS

According to [100], the aims of the 3GPP specifications are defined as follows.

- They should enable users to access a wide range of telecommunications services, including multimedia and high data rate services as well as types of services that are undefined today.

- They should facilitate the provision of a high quality of service similar to that provided by today's fixed networks.
- They should enable the design of small, easy to use, and low cost terminals with long talk and standby times.
- They should provide an efficient usage of network resources, especially the radio spectrum.

Existing 2G systems have standardized the complete set of teleservices, applications and supplementary services that they provide. This often requires a high standardization and implementation effort if new services are to be introduced. 3GPP takes a different approach by standardizing the service capabilities and not the services themselves. By definition, service capabilities consist of bearers defined by QoS parameters and the mechanisms needed to realize services. It is intended that these standardized capabilities should provide a defined platform that will enable the support of speech, video, multimedia, messaging, data, other teleservices, user applications and supplementary services and enable an unrestricted market for services.

Telecommunication services as defined by 3GPP are the communication capabilities that are available to the users [101]. Three different types of services are commonly distinguished. *Bearer services* are telecommunication services providing the capability of transmission of signals between access points. They are distinguished by a set of characteristics with certain requirements on QoS. QoS is the quality of a requested service as perceived by the customer and is always meant end-to-end. *Teleservices* provide the complete capability (including terminal equipment functions) for communication between users according to protocols established by agreements between network operators. Bearer services and teleservices are the so-called basic telecommunication services. A *supplementary service* modifies or supplements these basic telecommunication services. Supplementary services are used to complement and personalize the usage of bearer services and teleservices.

Bearer services. For a distinct bearer service, values are assigned to each characteristic. If networks between the two access points use different control mechanisms, the bearer services of each network throughout the communication link have to be translated at the network interfaces to realize an end-to-end bearer service. The bearer services are negotiable and can be used flexibly by applications. This negotiation can also take place during an active connection, e.g., to match the connection parameters to a changing environment. The characteristic parameters describing a bearer service are the following:

- Support of either connection-oriented or connectionless services
- One of the four possible traffic classes
  - **Conversational Class**: The conversational QoS class will be most widely used for classical real-time speech services. For this type of application the requirement characteristic is strictly determined by human

perception. Here, the most stringent requirement is that the delay stays below a certain value. Otherwise the quality of the transmission will be unacceptable to the user. Evaluations have been shown that the end-to-end delay (including the signal processing in the speech codecs) should be below 400 ms [75]. A second essential feature of this traffic class is that the time relation (variation) between information entities of the stream must be preserved. With Internet and multimedia, new applications like voice over IP or video telephony are emerging, which also require conversational QoS traffic class. Other examples are interactive games, two-way control telemetry, and telnet.

– **Streaming Class**: Because of the slow access for most Internet users today, streaming is an important technology for multimedia applications. With streaming, the data are transferred in such a way that they can be processed as a continuous steady stream, which implies that the time relation between information entities of the stream must be preserved in common with the conversational class. This allows the client browser to start displaying the data before the entire file has been received. Typical of streaming traffic is its highly asymmetric characteristic, which allows for higher delays and higher delay variations, since these variations can be smoothed out by buffering.

– **Interactive Class**: Interactive QoS class is applied for classical Web browsing. An end user requests data from a remote server. The data are transmitted (usually in packets) via the network to the user. While a certain round-trip limit must be kept for a satisfactory service, this delay constraint is much higher than in any real-time application like telephony. On the other hand the bit or packet error rate must be very low in data applications for otherwise the delay would strongly increase owing to some type of ARQ protocol.

– **Background Class**: A characteristic of background traffic is that the user or application does not expect the data within a certain time but expects it to be error free. Typical examples for applications using this type of traffic are email delivery, file download, or short message service (SMS). Usually the application is active in the background while the user is working with another application or is not actively using the terminal at all.

• Traffic characteristics
  – Point-to-Point
    * uni-directional
    * bi-directional (symmetric and asymmetric traffic)
  – Point-to-Multipoint
• Maximum transfer delay
• Delay variation
• Bit error rate
• Data rate

In [101] the valid range of values for each parameter is specified. The supported bit rates are given as:

- At least 144 kbps/s in rural outdoor radio environments.
- At least 384 kbps/s in urban/suburban outdoor radio environments.
- At least 2048 kbps/s in indoor/low range outdoor radio environments

Teleservices. Both standardized and non-standardized teleservices exist. Standardized teleservices are necessary because of the required interworking with other systems. Standardized teleservices are speech (standardized speech codec in the terminal and interworking units which allow calls to be received from or destined to users of existing networks), emergency calls, and SMS (point-to-point and cell broadcast). Since teleservices include the terminal functionality, they require the association of terminal and network capabilities.

6.3 Transport and Physical Channels

Data from higher layers are transmitted over so-called transport channels [102]. They are the services offered by layer 1 to the higher layers. A transport channel is defined by how and with what characteristics data are transferred over the air interface. The transport channels are mapped onto physical channels in layer 1. As described above, the MAC layer performs the mapping between the transport channels and the logical channels used in layers above the MAC layer.

We can distinguish two groups of transport channels. *Common channels* are used by all users in a cell. *Dedicated channels* are associated with a single MS. Table 6.2 shows the mapping of the existing transport channels onto the physical channels. There are several physical channels that are not visible to higher layers. Therefore, they have no corresponding transport channel mapped onto them. In the following, characteristics and functions of transport channels and physical channels are described together if there exists a one-to-one mapping between both channel types.

6.3.1 Dedicated Channel

The Dedicated Channel (DCH) is the only dedicated channel in UMTS. It is either an UL or a DL transport channel and carries all information from higher layers for a given user. The information transmitted on a DCH comprises data (e.g., video or speech) and control information (e.g., measurement reports or handover commands). Features of the DCH are fast power control and data rate change (every 10 ms) and the possibility of transmission over only part of the cell by means of an adaptive antenna array. The dedicated transport channel is mapped to the dedicated physical channel (DPCH). The

Table 6.2. Transport and physical channels and their mapping

Transport Channel		Physical Channel	
Dedicated Uplink Channels			
Dedicated Channel	DCH	Dedicated Physical Data Ch.	DPDCH
		Dedicated Physical Control Channel	DPCCH
Dedicated Downlink Channels			
Dedicated Channel	DCH	Dedicated Physical Data Ch.	DPDCH
		Dedicated Physical Control Channel	DPCCH
Common Uplink Channels			
Random Access Channel	RACH	Physical Random Access Channel	PRACH
Common Packet Channel	CPCH	Physical Common Packet Channel	PCPCH
Common Downlink Channels			
Broadcast Channel	BCH	Primary Common Control Physical Channel	P-CCPCH
Forward Access Channel	FACH	Secondary Common Control Physical Channel	S-CCPCH
Paging Channel	PCH	Secondary Common Control Physical Channel	S-CCPCH
Downlink Shared Channel	DSCH	Physical Downlink Shared Channel	PDSCH
		Primary Synchronization Ch.	P-SCH
		Secondary Synchronization Channel	S-SCH
		Primary Common Pilot Ch.	P-CPICH
		Secondary Common Pilot Channel	S-CPICH
		Acquisition Indication Ch.	AICH
		Page Indication Channel	PICH
		CPCH Status Indicator Ch.	CSICH
		CPCH Access Preamble Acquisition Indicator Ch.	AP-AICH
		CPCH Collision Detection/ Channel Assignment Indicator Channel	CD/CA-ICH

DPCH is different in downlink and uplink concerning its splitting in dedicated data and control channels (DPDCH and DPCCH). The DPDCH carries the dedicated transport channel which is offered to the MAC layer. The control information generated in the PHY is transmitted over the DPCCH.

Uplink DPDCH and DPCCH. In the uplink the data and control information are I/Q-multiplexed. In UMTS a modified form of QPSK, namely HPSK (Hybrid PSK) is used as modulation format (see section 6.4). The spread data bits are taken as the I-component of the baseband data while the spread control bits are modulated onto the Q-component. This type of modulation is called dual-channel QPSK. In conjunction with the complex scrambling (see 6.4.2) we yield HPSK. The structure of slots, radio frames and superframes is shown in Fig. 6.3 together with the I/Q-multiplexing. The DPDCH is multiplexed on the I-branch and the DPCCH

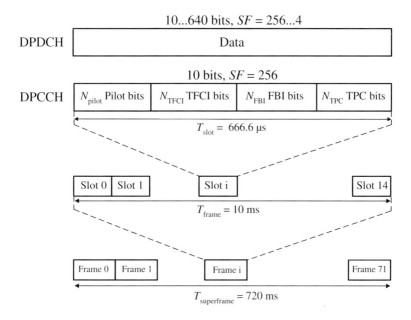

Fig. 6.3. Structure of superframes, radio frames and slots for the uplink DPCH

on the Q-branch. One slot always contains 2560 chips, each with a duration of $T_{chip} = 0.26042\,\mu s$ giving a total slot duration of $T_{slot} = 666.6\,\mu s$. Fifteen slots are grouped together, building a radio frame with a duration of $T_{frame} = 10\,ms$. A superframe of length $T_{superframe} = 720\,ms$ is made up of 72 radio frames.

The transmitted number of bits varies according to the spreading factor (SF) used. It should be noted that in UMTS SF is defined as *number of chips per symbol*. Owing to the I/Q-multiplexing in the uplink and the BPSK modulation of both branches, different SFs can be used for the I (data) and Q (control) branches and the number of symbols equal the number of bits. For the DPCCH the SF is always 256 to achieve the best protection against interference due to a high processing gain. Therefore, we have 10 bits per uplink DPCCH slot. For each DCH there must be one, but only one, DPCCH. According to Fig. 6.3, a DPCCH contains the following control information fields. The block of known *pilot bits* enables the RAKE receiver to estimate the mobile radio channel. In the *TFCI* (Transport Format Combination Indicator) field the transport format (spreading factor, number of bits in each DPCCH field, etc.) is encoded for the receiver to perform correct despreading and decoding. The use of the TFCI field is optional and is determined by the UTRAN. For fixed rate services no TFCI is transmitted, but its transmission is necessary for variable rate services, because in this case the data rate of the DPDCH can change after every radio frame (10 ms). The TFCI bits of a whole radio frame are necessary to decode the transport format. If the TFCI is not decoded correctly, the whole frame is lost, but no other frames are affected. The third field in the DPCCH contains the *FBI* (Feedback Information) bits. This feedback path from the MS to the BS is necessary to enable techniques such as closed loop mode transmit diversity and site selection diversity transmission. Details of both techniques and the associated FBI field use can be found in [103] and partly in section 6.6.5. The last field in the DPCCH slot format contains the *TPC* (Transmit Power Control) bits. As described in section 4.3, fast and accurate power control is vital to any CDMA system. Owing to the TPC field in every slot, the transmit power of the BS is updated at a rate of 1.5 kHz. The BS must support a stepsize for a change of the transmit power in every slot of 1 dB, the support of 0.5 dB is optional [104] (see also section 6.6.2).

A slot of the uplink DPDCH carries only data. To support a variety of data rates, different SFs can be used. They range from 256 to 4 resulting in 10 to 640 bits per slot, which gives a physical bit rate of 15 to 960 kbit/s over the air interface, i.e., including error correction coding. In Table 6.3 the achievable user data rates on the UL DPDCH are listed if coding with rate 1/2 is assumed. The details of the different slot formats are given in Table

Table 6.3. UL DPDCH data rates

DPDCH SF	DPDCH channel bit rate (kbps)	Maximum user data rate[1] (kbps)
256	15	7.5
128	30	15
64	60	30
32	120	60
16	240	120
8	480	240
4	960	480
4 with 6 parallel codes	5740	2300

[1] Approximate data rate if a coding rate of 1/2 is assumed.

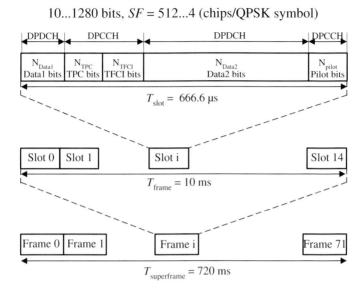

10...1280 bits, $SF = 512...4$ (chips/QPSK symbol)

Fig. 6.4. Structure of superframes, radio frames and slots for the downlink DPCH

C.5 in Appendix C.3. Since UMTS supports data rates for one user of up to 2 Mbit/s, more than one DPDCH can be assigned to each user. The number of DPDCHs is specified from 0 to 6. They are multiplexed alternately to the I- and Q-branch. However, a DCH should be transmitted over only one DPDCH if possible (from the point of view of the required data rate, the requested link quality, and the actual transmission conditions), because the more channels that are transmitted in parallel, the higher the peak-to-average ratio (crest factor, see section 2.4.3) of the RF-signal becomes. A high crest factor in turn requires a high backoff of the power amplifier, which lowers the achievable efficiency.

If control information and data is time-multiplexed, pulsed transmission over the air interface could occur in the case of burst-type data traffic (e.g., packet data or speech transmission with discontinuous transmission (DTX) due to voice activity detection (VAD)). Due to electromagnetic compatibility (EMC) effects, pulsed transmission could cause audible interference to audio equipment which is located close to the MS [76]. This is prevented by the I/Q-multiplexing of data and control information.

Downlink DPDCH and DPCCH. The main difference between UL and DL DPCHs is their multiplexing of DPDCH and DPCCH. For the DL, this multiplexing is performed in time. Furthermore, the SF is not changed after every radio frame. Any necessary data rate variation is realized using either rate matching or DTX. This produces no audible interference due to the continuous reception of the common DL channels. Figure 6.4 shows the DL

Table 6.4. DL DPDCH data rates

Spreading factor	Channel symbol rate (ksps)	Channel bit rate (kbps)	DPDCH bit rate range (kbps)	Maximum user data rate[1] (kbps)
512	7.5	15	3–6	1–3
256	15	30	12–24	6–12
128	30	60	42–51	20–24
64	60	120	90	45
32	120	240	210	105
16	240	480	432	215
8	480	960	912	456
4	960	1920	1872	936
4 with 3 parallel codes	2880	5760	5616	2300

[1] Approximate data rate if a coding rate of 1/2 is assumed.

slot, frame and superframe structure. Again, we have 2560 chips per slot. The radio frame and superframe structure are equal to the UL, while the slot structure differs. As in the UL, the achievable data rate depends on the SF. Because of the use of QPSK (HPSK) and the time-multiplexing of control information and data, each (QPSK-)symbol is equal to two bits. Therefore, with a SF ranging from 512 to 4, the number of bits per slot is between 10 and 1280, resulting in an overall data rate of 15 to 1920 kbps. This data rate includes the control information, which comprises the pilot bits, the TPC bits and the TFCI bits, as is the case in the UL. Table 6.4 lists the achievable data rates on the DL DPDCH. Details of the different slot formats are listed in Table C.4 in Appendix C.2. In the DL three parallel DPCHs are sufficient to achieve a user data rate of 2 Mbps.

6.3.2 Random Access Channel

The random access transport channel (RACH), which is mapped to the physical random access channel (PRACH), is a common transport channel in the UL. It is used to transmit control information, e.g., a request for setting up a connection, to register the MS after power-on, or to perform a location update. Packet data can also be carried by the RACH if the traffic volume is low. The random access scheme is based on the slotted ALOHA protocol [102], where the MS can start its transmission at the beginning of specified access slots. A total of 15 access slots per two frames are specified and they are 5120 chips apart. The RACH uses open loop power control, it must be received from the entire cell, and there is certain collision risk. The physical format of the RACH is the same as for the UL DPCH (see Fig. 6.3) but only pilot and TFCI bits are transmitted as control information. Specific to the PRACH are one or several *preambles* with a length of 4096 chips each. They consist of 256 repetitions of a signature of 16 chips. After detection of the preamble in the BS and acknowledgement via the acquisition indication channel (AICH) (section 6.3.10), the 10 or 20 ms long *message part* is transmitted. The data part of the message supports bit rates from 15 to 120 kbit/s (SF from 256 to 32). The physical random access procedure is described in section 6.6.3 and in more detail in [103].

6.3.3 Common Packet Channel

The physical common packet channel (PCPCH) carries the common packet channel (CPCH), a UL transport channel. It is a contention based random access channel with a physical layer-based collision detection mechanism (CSMA-CD, carrier sense multiple access with collision detection) used for transmission of packet-type data traffic. The access slot timing and structure, as well as the general physical format, are identical to those of the RACH. Only the contents of the control part differs. Another difference to the RACH is the use of fast power control. For this, a DL DPCCH is associated with the PCPCH, while higher layer DL signaling is performed via the forward access channel (FACH, see section 6.3.5). The CPCH access transmission consists of one or several access preambles 4096 chips long, one collision detection preamble (CD-P) 4096 chips long, an DPCCH power control preamble (PC-P) that is either 0 slots or 8 slots in length, and a message of variable length, which is an integer multiple of 10 ms. A more detailed description can be found in section 6.6.4 and in [103].

6.3.4 Broadcast Channel

The broadcast channel (BCH), a common DL channel, is mapped onto the primary common control physical channel (P-CCPCH). Via this channel, cell- and system-specific information is transmitted to all MSs in a cell. This information usually contains data about available codes and the timing, e.g., of the access slots. A MS must decode the BCH to register in the cell. For this reason the P-CCPCH is usually transmitted with high power to cover the whole cell. A P-CCPCH slot contains only 18 data bits (no TPC, TFCI, or pilot bits) with a SF of 256. During the first 256 chips no BCH data are transmitted. In this interval the primary and secondary synchronization channels (P-SCH and S-SCH) are active (see section 6.3.8).

6.3.5 Forward Access Channel

The forward access channel (FACH) is a DL transport channel to be transmitted over the entire cell or over a part of the cell using, e.g., beam-forming antennas. The FACH transmits control information to MSs that are known by the UTRAN to reside in a given cell. Short user data packets can also be transmitted via the FACH. If more than one FACH is active in one cell, at least one must have such a low data rate that all MSs in the cell are able to receive it. The FACH is mapped onto the secondary common control physical channel (S-CCPCH) that also carries the paging channel (PCH, see section 6.3.6). FACH and PCH can be mapped either to the same or to separate S-CCPCHs. If FACH and PCH are mapped to the same S-CCPCH, they can be mapped to the same frame. The reason for multiplexing both transport channels to the same S-CCPCH is the power budget of the BS. Both channels must be transmitted at full power to reach all MSs. If they are transmitted at different physical channels, the interference would be higher as in the case of multiplexing them to one physical channel. The slot structure of a S-CCPCH is similar to that of the downlink DPCH but without the first data field and the TPC bits. Therefore, no inner-loop power control is performed on this channel. The data rate on the S-CCPCH depends on the SF ranging from 256 to 4. Bit rates from 30 to 1920 kbps are possible [102].

6.3.6 Paging Channel

The paging channel (PCH), again a DL transport channel, carries control information for the paging procedure to the MS in case the system does not know its location cell. Paging is needed if the UTRAN wants to start communication with a MS. Therefore, the PCH must be transmitted over the entire cell.

6.3.7 Downlink Shared Channel

As the name suggests, the downlink shared channel (DSCH) is a transport channel to transmit dedicated user data and/or control information and can be shared by several MSs. The introduction of this channel type provides the transmission of packet traffic with high peak data rates but low activity cycle without allocating too much of the channelization codes. The DSCH is mapped to the physical downlink shared channel (PDSCH). The sharing mechanism is code multiplexing. Each shared channel is always associated with one or several dedicated channels. This is valid for transport and physical channels. Each PDSCH radio frame is associated with one downlink DPCH to obtain the necessary power control and decoding information. A slot of the PDSCH contains only data with rates from 30 to 1920 kbps (i.e., the SF ranges from 4 to 256), which can change from frame to frame.

6.3.8 Synchronization Channel

The synchronization channel (SCH) is a physical channel that is not visible to higher layers. It consists of two sub-channels, the primary and secondary

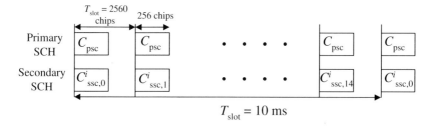

Fig. 6.5. Synchronization channel structure

synchronization channel (P-SCH and S-SCH), and is necessary for the cell search procedure.

The P-SCH consists of a spreading sequence of 256 chips called the primary synchronization code (PSC) that is denoted by C_{psc}. The sequence is transmitted at the beginning of every slot and is the same for every cell in the system. The transmission takes place during the idle period of the P-CCPCH that carries the BCH (section 6.3.4).

The S-SCH is transmitted in parallel to the P-SCH. Instead of repeating the same code in very slot, the S-SCH is composed of a sequence of 15 different codes each 256 chips long. These codes are termed the secondary synchronization codes (SSCs), $C^i_{ssc,k}$. In total there are 16 different SSCs. The sequence of 15 of these 16 SSCs is is repeated every radio frame (the index k ranges from 0 to 14 and is the slot number of the radio frame). There are 64 different sequences of SSCs specified and each sequence, denoted by i, identifies the scrambling code group (see section 6.4.2) to which the BS belongs. This scrambling code group, encoded by the sequence of the SSCs, has to be determined by the MS during synchronization. Figure 6.5 illustrates the synchronization channel structure. The synchronization procedure performed during cell search is further described in section 6.6.1.

6.3.9 Common Pilot Channel

Similar to the SCH, the common pilot channel (CPICH) is a DL physical channel which is not visible to higher layers. It consists of two sub-channels, the primary and secondary common pilot channel (P-CPICH and S-CPICH). The purpose of the CPICH is to aid the channel estimation in the MS for the DCHs and to provide the phase reference for the channel estimation for common channels that are not associated with dedicated channels. A predefined symbol sequence with a fixed bit rate of 30 kbps (SF of 256) is transmitted on the CPICH. There is always one P-CPICH in each cell with the fixed channelization code $c_{ch,256,0}$ and scrambled with the cell's primary scrambling code (see section 6.4). The P-CPICH is also used to perform the measurements necessary for the handover procedure to other cells in the system. Therefore, the adjustment of the power level of the P-CPICH can change the number of users in a cell. If the P-CPICH power is decreased, some users will hand over to neighboring cells with higher P-CPICH power.

The S-CPICH can use any of the 256 chip long channelization codes and is scrambled either by the primary or a secondary scrambling code. In a cell there may be zero, one or several S-CPICHs. They are used primarily with narrow antenna beams directed to areas with high traffic, because in this case the mobile radio channel is different from the channel that a signal transmitted over an omnidirectional antenna experiences. They can be the reference for the S-CCPCH and the DL DPCH. For details on the use of the CPICH with transmit diversity, see section 6.6.5 and [102].

6.3.10 Acquisition Indication Channel

The acquisition indication channel (AICH), a DL physical channel without any transport channel mapped onto it, is used in conjunction with the RACH/PRACH (section 6.3.2). If the BS detects a preamble on the RACH it sends its acknowledgement via the acquisition indicators (AIs) which correspond to the signature of the PRACH. Since the MS sends its RACH message part only after detecting the AI, it must be received in the whole cell. Therefore, the AICH is typically transmitted at a high power level without power control. Because the physical layer directly controls the AICH, it can respond rapidly to a random access trial of the MS.

6.3.11 Paging Indicator Channel

The paging indicator channel (PICH) is very similar to the AICH. It is a DL physical channel, not visible to higher layers, and is used always in connection with the PCH (mapped to the S-CCPCH). The PICH uses a fixed rate of 30 kbps (SF of 256) to carry the paging indicators (PIs). A PICH radio frame consists of 300 bits of which 288 bits are used to carry the PIs. The remaining 12 bits are undefined. In each PICH frame, 18, 36, 72, or 144 PIs are transmitted. If a PI is set to "1" it is an indication that the MSs associated with this PI should read the corresponding frame of the associated S-CCPCH. How often a MS listens to the PICH determines the stand-by battery life and the response time to a paging call. A compromise between these two contrary requirements must be found.

6.3.12 CPCH Status Indicator Channel

Via the CPCH status indicator channel (CSICH) the status information of the different CPCHs in a cell are transmitted by the BS and can be monitored by the MSs. The CSICH is a fixed rate (SF = 256) physical channel and has a 40 ms frame structure consisting of 15 so-called access slots. The values of the status indicators transmitted in the access slots are determined by higher layers in the UTRAN.

6.3.13 CPCH Access Preamble Acquisition Indicator Channel

Similar to the AICH, the CPCH access preamble acquisition indicator channel (AP-AICH) carries the acquisition indication for access attempts via the CPCH. The AP acquisition indicator (API) corresponds to AP signatures transmitted by the MS. AP-AICH and AICH can use the same channelization codes.

6.3.14 CPCH Collision Detection/Channel Assignment Indicator Channel

To detect as well as reduce collisions, appropriate information is transmitted from the BS to the MS on the collision detection/channel assignment indicator channel (CD/CA-ICH). An explicit channel assignment (CA) for the message part of the CPCH is optional. If CA is not active, the CD/CA-ICH carries the Collision Detection Indicator (CDI) only. Otherwise CDI and the Channel Assignment Indicator (CAI) are transmitted simultaneously. CD/CA-ICH and AP-AICH can use the same channelization codes.

6.4 Spreading and Modulation

The spreading applied to the physical channels increases the bandwidth, thus generating the wideband direct sequence SS signal (see section 3.2). In UMTS, this operation is called channelization, because the use of orthogonal codes makes it possible to separate the different channels from a single source in the receiver. An additional operation, the *scrambling*, is applied in UMTS after the channelization. This is to separate different BSs or MSs from each other and does not change the bandwidth. Both operations will be described in the following according to [106].

6.4.1 Channelization

The channelization operation spreads the bandwidth of the data signal by converting each bit/symbol into a number of so-called chips and ensures that different physical channels from one source (MS or BS) can be separated in the receiver. In UMTS, the bandwidth spreading must be variable to support the different data rates. Codes that fulfill these requirements and are used in UMTS for channelization are the orthogonal variable spreading factor (OVSF) codes, originally proposed in [107]. The codes are taken from a code tree, shown in Fig. 6.6. The channelization codes are identified by $c_{ch,SF,k}$ with the spreading factor SF and the code number $0 \leq k \leq SF - 1$. Therefore, the available number of codes with a certain SF is equal to the SF. The numbering of the codes in the code tree is performed from the top down,

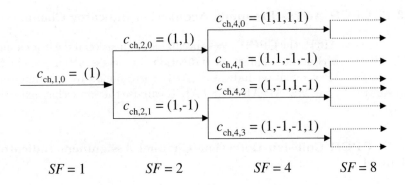

Fig. 6.6. Orthogonal Variable Spreading Factor (OVSF) code tree

beginning with code number $k = 0$, as is shown in Fig. 6.6. The codes are generated using a simple algorithm. To build the code of two new branches with SF $= L$, take the code of the lower branch (SF $= L - 1$) and repeat it for the first branch with SF $= L$. For the second branch change the sign of the repeated code. This algorithm is similar to the recursive generation of the Walsh codes by means of the Hadamard matrices. In fact, the OVSF codes are identical to the Walsh codes.

For choosing codes from the tree, some restrictions apply to ensure mutual orthogonality. First, all codes with the same SF are orthogonal to each other. Two codes with different SFs are not orthogonal if the code with the higher SF can be constructed out of the code with the lower SF. This is equal to the fact that the code with lower SF lies "on the way" from the code with the higher SF down to the root of the code tree.

In UMTS, the OVSF codes are used for UL and DL. SFs from 4 to 256 (UL) or 512 (DL) are used. The SF is defined as chips per symbol and channel. Due to I/Q-multiplexing of DPDCH and DPCCH in the UL, each channel has binary symbols. In DL, we have time-multiplexing of data and control DPCHs leading to quaternary symbols (QPSK).

Uplink Spreading. The UL spreading structure, including channelization and scrambling, is shown in Fig. 6.7. As already described in section 6.3.1, each UL DCH consists of always one DPCCH, which is mapped to the Q-branch, and of zero up to a maximum of six DPDCHs, depending on the needed data rate and propagation conditions. The DPDCHs are alternately added to the I- and Q-branch, starting with DPDCH1 at the I-branch. The bit streams of the channels are represented by real-valued sequences, where the binary values $\{0, 1\}$ are mapped to the real values $\{1, -1\}$ before spreading. Each channel is independently spread to the chip rate with its OVSF channelization code, which is denoted c_c for the DPCCH and $c_{d,n}$ for the DPDCHs with $0 \leq n \leq 6$. The DPCCH is always spread by the code $c_c = c_{ch,256,0}$ (SF

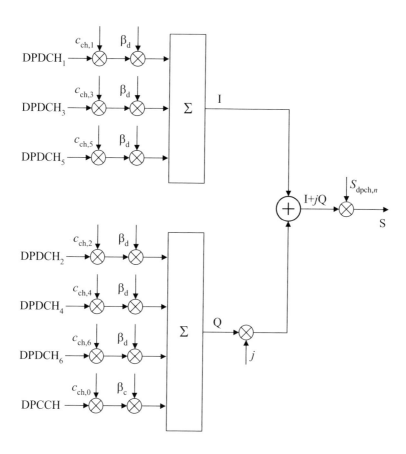

Fig. 6.7. Uplink spreading and scrambling

of 256 and uppermost branch in the OVSF code tree according to Fig. 6.6). For the DPDCHs we have to distinguish between two cases [106]:

- When *only one* DPDCH is to be transmitted, DPDCH1 is spread by code $c_{d,1} = c_{ch,SF,k}$ where SF is the spreading factor of DPDCH1 and $k = SF/4$.
- When *more than one* DPDCH is to be transmitted, all DPDCHs have a spreading factor of 4. Then DPDCHn is spread by the the the code $c_{d,n} = c_{ch,4,k}$, where $k = 1$ if $n \in \{1, 2\}$, $k = 3$ if $n \in \{3, 4\}$, and $k = 2$ if $n \in \{5, 6\}$. Because of the I/Q multiplexing of the multiple DPDCHs one code can be used for two channels if they are not both mapped on the same branch.

After the spreading with the channelization codes, we still have real-valued sequences taking values from $\{\pm 1\}$. These sequences are weighted by gain factors, β_c for the DPCCH and β_d for all DPDCH's. At least one of the values β_c and β_d must have the amplitude 1. The gain factors are quantized into 4-bit words given in Table C.6 in Appendix C.4. After the weighting with

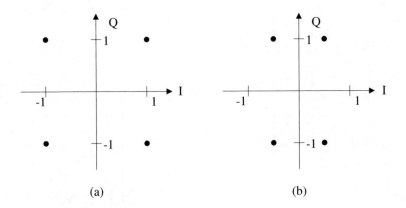

Fig. 6.8. Constellation diagrams for two channels (DPDCH and DPCCH) with (a) equal and (b) different gain factors ($\beta_c = 1$ and $\beta_d = 0.5$)

the gain factors the streams of real-valued chips on the I- and Q-branches are summed up and the value in the Q-branch is multiplied with j and added to the value of the I-branch, yielding a complex-valued chip sequence (I+jQ).

The weighting with the gain factors β_c and β_d makes it possible to assign different powers to different channels. This is beneficial from the MAI point-of-view. Depending on the required QoS and bit rate, higher power in either the data or the control channel could be useful. Transmitting all channels with their lowest possible power (which are usually different) will result in minimum MAI. From the power amplifier (PA) point-of-view, this weighting deteriorates the performance. In general, for a linear modulation format such as QPSK, a highly linear PA is required, which usually has a poor efficiency. The linearity requirement is still further increased if the channels have different weighting. This can be seen from the constellation diagrams shown in Fig. 6.8. If the weighting factors are introduced (Fig. 6.8b) the peak power is reduced only slightly but the reduction of the average power is more pronounced, leading to a higher crest factor. This is because transitions from, e.g., $\{+0.5, +1\}$ to $\{+0.5, -1\}$ happen at lower amplitudes as in the case of a quadratic constellation. We will see that the complex scrambling described in section 6.4.2 copes with this problem and leads to a crest factor of the transmit signal that is nearly independent of the channel weighting.

If multiple data channels are transmitted (multi-code transmission, see Fig. 6.9) the situation gets worse. New points in the constellation diagram appear which correspond to higher signal amplitude than for conventional QPSK (Fig. 6.9b). If gain factors are introduced in the multi-code case, the crest factor increases further (Fig. 6.9c).

So far only the channelization part of the UL DCH has been described. For the two common UL channels (RACH/PRACH and CPCH/PCPCH) the

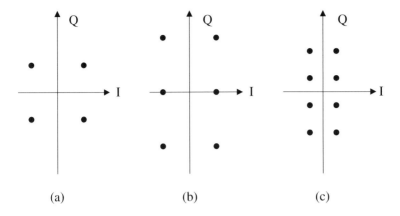

Fig. 6.9. Constellation diagram for (a) one DPCCH and one DPDCH with equal gains and one DPCCH and two DPDCH's with (b) equal gains and (c) $\beta_d = 0.5$

procedure is the same. Concerning the specific channelization code allocation we refer to [106].

Downlink Spreading. The DL spreading structure including channelization and scrambling is shown in Fig. 6.10. Due to the time multiplexing of data and control channel conventional QPSK modulation is used in the DL. Figure 6.10 applies for all DL channels except for the SCH. The data symbols take the values out of $\{-1, 0, +1\}$ on each channel with 0 for the case of discontinuous transmission (see section 6.5). After converting each pair of two consecutive bits into a QPSK symbol in the serial-to-parallel converter and mapping it to the I and Q branch, the branches are then spread to the chip rate by the same real-valued channelization code $c_{ch,SF,m}$. After summing this gives a complex-valued chip sequence, which is then scrambled (see section 6.4.2).

The channelization codes are the same OVSF codes as in the UL. The P-CPICH always uses the code $c_{ch,256,0}$ and the code for the P-CCPCH is fixed to $c_{ch,256,1}$. All other channelization codes are assigned by the UTRAN. Some minor restrictions to code allocation are specified in [106]. As can be seen from Table 6.4, for one user, three of four codes with a SF of 4 have to be allocated to achieve a net data rate of 2 Mbps. If more than one user requests the maximum data rate in the FDD mode, different scrambling codes have to be allocated for the different users. As described in section 6.4.2, each cell is associated with one primary and 15 secondary scrambling codes. The different secondary scrambling codes can be used, e.g., for different DL DPCHs. However, the most probable solution in the case that one or more users request a data rate of 2 Mbps is a transmission in the TDD mode (see chapter 7).

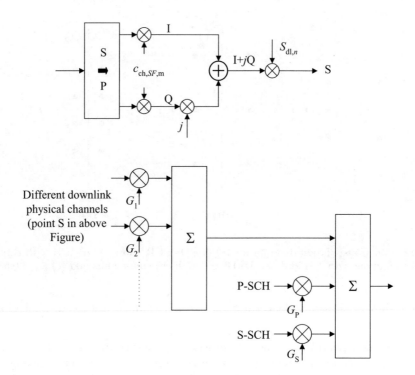

Fig. 6.10. Downlink spreading and scrambling

After scrambling, each channel is weighted by a weight factor G_i and they are complex added. The primary and secondary SCH are not scrambled. They have their associated weighting factors, G_p and G_s.

6.4.2 Scrambling

Scrambling in UMTS is the multiplication of the complex-valued spread chip sequence with a complex-valued scrambling code. This scrambling operation does not increase the bandwidth of the signal further. While the channelization code per definition separates the different channels of one user in the UL and the different dedicated and common channels in the DL, respectively, the scrambling codes separate the MSs in the UL and the BSs or BS sectors in the DL, i.e., the sum of all channels of a BS is scrambled with one scrambling code in the DL and the sum of all channels of a MS is scrambled with one scrambling code in the UL.

Uplink. In the UL the complex-valued DPCCH/DPDCH sequence is scrambled with either a long or a short scrambling code $S_{dpch,n}$. In total 2^{24} long and 2^{24} short scrambling code are available. The *long scrambling codes* are

Table 6.5. Mapping from $z_n(i)$ to $c_{\text{short},1,n}(i)$ and $c_{\text{short},2,n}(i)$

$z_n(i)$	$c_{\text{short},1,n}(i)$	$c_{\text{short},2,n}(i)$
0	+1	+1
1	−1	+1
2	−1	−1
3	+1	−1

constructed as follows [106]. The chip-wise modulo 2 addition of two m-sequences, defined by two generator polynomials of degree 25, gives a Gold code $z_n(i)$ (see Appendix C.5.1). Now two real-valued long sequences are defined by mapping the binary values 0 and 1 of the Gold sequence to +1 and −1 to yield $c_{\text{long},1,n}$, and performing the same mapping but with the original Gold sequence shifted by 16,777,232 chips to get $c_{\text{long},2,n}$. Here n is the scrambling sequence number ranging from 0 to $2^{24} - 1$. Its binary representation specifies the initial shift register values for the generation of one of the two m-sequences. The complex-valued scrambling sequence is now defined as

$$C_{\text{long},n}(i) = c_{\text{long},1,n}(i) \left[1 + j(-1)^i c_{\text{long},2,n} \left(2 \lfloor i/2 \rfloor \right) \right] \tag{6.1}$$

with $i = 0, 1, ..., 2^{25} - 2$ and $\lfloor \cdot \rfloor$ denoting the nearest lower integer. Now the first 38,400 values (which correspond to one frame) of the long sequence are taken for the final nth scrambling code for the UL DCH.

$$S_{\text{dpch},n}(i) = C_{\text{long},n}(i), \qquad i = 0, 1, ..., 38399. \tag{6.2}$$

The scrambling code is repeated for every frame. The complex valued sequence $S_{\text{dpch},n}(i)$ takes on values out of $\{\pm 1 \pm j\}$.

Besides the use of the long scrambling codes, UMTS supports *short scrambling codes*. They are used if advanced interference cancellation receivers or multiuser detectors are applied in the BS. The short scrambling codes make the implementation of these receiver structures easier due to the reduced computational load. To construct the short codes, two binary and one quaternary sequences are modulo 4 added. The resulting sequence $z_n(i)$ takes on values out of $\{0, 1, 2, 3\}$. It is of length 255 and is extended to length 256 by the extension of one chip with value 0. For further details, see Appendix C.5.1. The sequence values are mapped to $c_{\text{short},1,n}(i)$ and $c_{\text{short},2,n}(i)$ according to the Table 6.5. As with the long scrambling codes, the complex-valued scrambling sequence is defined as

$$C_{\text{short},n}(i) = \tag{6.3}$$

$$c_{\text{short},1,n}(i \bmod 256) \left[1 + j(-1)^i c_{\text{short},2,n} \left(2 \lfloor (i \bmod 256)/2 \rfloor \right) \right]$$

with $\lfloor \cdot \rfloor$ denoting the nearest lower integer. The nth scrambling code for the UL DCH is now

$$S_{\text{dpch},n}(i) = C_{\text{short},n}(i), \qquad i = 0, 1, ..., 38,399, \tag{6.4}$$

if short scrambling codes are used. The scrambling codes, both long and short, are applied aligned with the radio frames, i.e., the first scrambling chip corresponds to the beginning of a radio frame.

The message part of the PRACH is scrambled with one of 8192 different, 10 ms long scrambling codes $S_{\text{r-msg},n}$. They are defined as

$$S_{\text{r-msg},n}(i) = C_{\text{long},n}(i + 4096) \tag{6.5}$$

with $i = 0, 1, ..., 38, 399$, $n = 0, 1, ..., 8191$, and $C_{\text{long},n}$ from (6.1). The scrambling code for the PRACH preamble part is defined as

$$S_{\text{r-pre},n}(i) = c_{\text{long},1,n}(i) \tag{6.6}$$

with $i = 0, 1, ..., 4095$, $n = 0, 1, ..., 8191$, and $c_{\text{long},1,n}$ from the definition of the complex-valued spreading sequence in (6.1). For one PRACH, the same code number is used for the scrambling codes of both preamble and message part.

The PCPCH message part is scrambled with either a long or a short cell-specific scrambling code. There are 32,768 different PCPCH scrambling codes defined in the system and 64 UL scrambling codes defined in each cell. For the long scrambling codes we have

$$S_{\text{c-msg},n}(i) = C_{\text{long},n}(i) \tag{6.7}$$

with $i = 0, 1, ..., 38, 399$, $n = 8192, 8193..., 40, 959$, and $C_{\text{long},n}$ defined in (6.1). The definition in the case of short scrambling codes is

$$S_{\text{c-msg},n}(i) = C_{\text{short},n}(i) \tag{6.8}$$

with $i = 0, 1, ..., 38, 399$, $n = 8192, 8193..., 40, 959$, and $C_{\text{short},n}$ defined in (6.3). The 32,768 PCPCH scrambling codes are divided into 512 groups with 64 codes each. The group of PCPCH preamble scrambling codes in a cell corresponds to the primary scrambling code used in the DL of the cell. The kth PCPCH scrambling code within the cell with DL primary scrambling code number m, $k = 16, 17, ..., 79$ and $m = 0, 1, 2, ..., 511$ is $S_{\text{c-msg},n}$ as defined above with $n = 64m + k + 8176$.

In the following, we will review the effect of the complex scrambling. We start with the constellation diagrams shown in Fig. 6.11. Due to unequal gain factors the constellation is stretched (Fig. 6.11b) compared to the conventional QPSK constellation (Fig. 6.11a). The chip-wise multiplication of the complex-valued chip sequence $I_{\text{chip}} + jQ_{\text{chip}}$ with the complex-valued scrambling sequence $S_{\text{dpch},n}$ rotates the constellation by $\varphi_{\text{dpch}} \in \{-135, -45, +45, +135\}$, since $S_{\text{dpch},n} = \pm 1 \pm j$,

$$S = (I_{\text{chip}} + jQ_{\text{chip}}) \cdot (S_{\text{I,dpch}} + jS_{\text{Q,dpch}}) \tag{6.9}$$

$$= A_{\text{chip}} A_{\text{dpch}} \cdot \exp\{\varphi_{\text{chip}} + \varphi_{\text{dpch}}\}. \tag{6.10}$$

For the case of the original QPSK constellation, the scrambling results in a constellation that is shifted by $\pi/4$ (Fig. 6.12a). For the stretched constellation with different powers in I- and Q-branch the $\pi/4$ rotation leads to eight

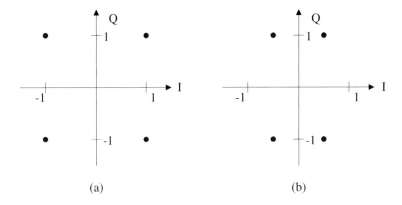

Fig. 6.11. Constellation diagrams for two channels (DPDCH and DPCCH) with (a) equal and (b) different gain factors ($\beta_c = 1$ and $\beta_d = 0.5$)

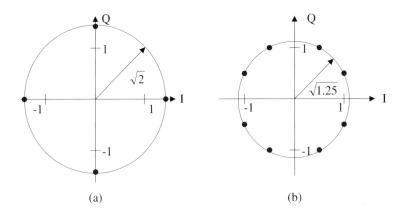

Fig. 6.12. Constellation diagrams for two channels (DPDCH and DPCCH) with (a) equal and (b) different gain factors ($\beta_c = 1$ and $\beta_d = 0.5$) after complex scrambling

instead of four different points in the constellation diagram and to equal power distribution between both branches, as can be seen from Fig. 6.12b.

As we will see, the complex scrambling together with the special construction mechanism of the scrambling sequence performs a reduction of the average number of transitions of the signal amplitude through zero (due to a phase shift of π between two chip values). This results in a reduction of the crest factor of the signal. Together with the dual-channel QPSK, the resulting modulation scheme is called HPSK [108]. To explain this in detail, we first recall the definition of the complex-valued UL scrambling sequence.

$$C_{\text{long},n}(i) = c_{\text{long},1,n}(i) \left[1 + j\,(-1)^i\, c_{\text{long},2,n}\left(2\left\lfloor i/2 \right\rfloor\right) \right] \tag{6.11}$$

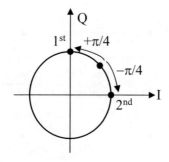

Fig. 6.13. Rotation of two equal chip values $(+1 + j)$ by $+\pi/4$ and $-\pi/4$ due to the multiplication with the Walsh rotator. The amplitude scaling is neglected for simplicity

with $\lfloor \cdot \rfloor$ denoting the nearest lower integer. We start with neglecting the two Gold sequences by setting them both to 1 $(c_{\text{long},1,n}(i) = c_{\text{long},2,n}(i) = 1, \forall i)$. What remains is the so-called Walsh rotator $1 + j(-1)^i$ [108], from which the sequence $1 + j, 1 - j, 1 + j, 1 - j, 1 + j, \ldots$ is constructed. If we now assume two consecutive chips to be equal, their constellation points after scrambling have always a phase difference of $\pi/2$. This can easily be seen, because the multiplication of the first chip value with $1 + j$ adds a phase shift of $+\pi/4$ while $1 - j$ rotates the second chip (with equal value) by $-\pi/4$ (see Fig. 6.13, where the amplitude scaling has been neglected for simplicity). The $\pi/2$ phase shift between two consecutive chip values ensures that a transition through zero does not occur. It is, however, necessary to assume pairs of chips with identical values. In UMTS, this is assured at least for the case of just one DPDCH. On this occasion the used channelization code number k is specified according to $k = \text{SF}/4$ (section 6.4.1). It can be easily verified, with the help of Fig. 6.6, that these codes consist of alternating pairs of chips with the same value (i.e. $+1 + 1 - 1 - 1 + 1 + 1 \ldots$) for all values of SF.

HPSK does not completely eliminate the transitions through zero because such a transition is prevented only between a pair of consecutive chips. Between two pairs of chips a transition through zero may occur, but the overall probability of zero-crossings is reduced, which also reduces the crest factor. An additional effect of HPSK is the elimination of a zero phase shift at the same transitions where the zero-crossings are prevented. A zero phase shift occurs when two consecutive chip values are equal. Because of the scrambling with the Walsh rotator, a $\pi/2$ phase shift between the two chip values appears as is described above. This is advantageous since a zero phase shift transition causes a high peak amplitude, as shown in Fig. 6.14, owing to the transmit filtering by means of a root-raised cosine filter (see section 6.4.4). Again, the zero phase shift transitions are not completely eliminated, but their probability is lowered, resulting in a reduced crest factor.

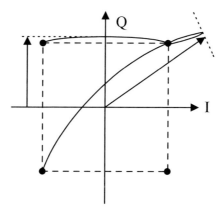

Fig. 6.14. Signal amplitude trajectory for transitions with phase shifts of zero and $\pi/2$

In UMTS the scrambling codes in the UL are needed to differentiate between the MSs. Therefore, the Walsh rotator $(1 + j, 1 - j)$ is multiplied with the Gold sequence. This multiplication does not change the $\pi/2$ phase shift property of the Walsh rotator, since both real and imaginary parts of the Walsh rotator are multiplied with either $+1$ or -1. This is shown in Fig. 6.15 for all four possible combinations of two consecutive chip values of $c_{\mathrm{long},1,n}$.

What remains to be described in (6.11) is the multiplication of the imaginary part of the Walsh rotator $(j\,(-1)^i)$ with the second Gold sequence $c_{\mathrm{long},2,n}\,(2\,\lfloor i/2 \rfloor)$. If we take a closer look at the argument of this sequence, we find that for a linear increase of i $(0, 1, 2, 3, 4, ..., 2n - 1, 2n, 2n + 1, ...)$ the argument takes on the values $0, 0, 2, 2, 4, ..., 2n - 2, 2n - 2, 2n, 2n, ...$, i.e., a kind of decimation is performed. Therefore, the factor $j(-1)^i$ in (6.11) is multiplied with the same sequence value (±1) of $c_{\mathrm{long},2,n}$ for two consecutive chips. This not only ensures the alternating sign of the imaginary part of the two consecutive values of the Walsh rotator but also randomizes the direction of the phase rotations. This effect is illustrated in Fig. 6.16.

From simulations the effect of the complex scrambling has been determined. For example, if we take the 768 kbps UL reference measurement channel [105], which is composed of 2 DPDCHs with 960 kbps each, the crest factor reduces from 3.909 in the unscrambled case to 3.458 with complex scrambling. In Fig. 6.17, the constellation diagram and the CDF (cumulative distribution function) of the signal amplitude is plotted for the unscrambled case. If scrambling is applied, a clear reduction of the probability of low signal amplitude values can be observed in Fig. 6.18, which results in the lower crest factor.

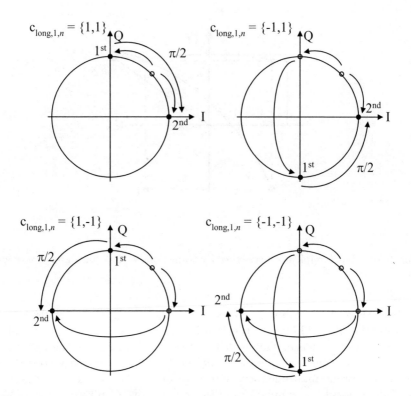

Fig. 6.15. Effect of multiplication of the Walsh rotator with the Gold sequence $c_{\text{long},1,n}$

Downlink. In the DL, the sequences of real-valued chips on the I and Q branch (see upper part of Fig. 6.10) are treated as a single complex-valued sequence of chips. This sequence of chips is scrambled by a complex chip-wise multiplication with a complex-valued scrambling code $S_{\text{dl},n}$. A total number of $2^{18} - 1 = 262,143$ scrambling codes can be generated, but not all of them are used. The scrambling codes are divided into 512 sets each consisting of a primary scrambling code and 15 secondary scrambling codes giving a total of 8192 codes. The numbering rules are as follows. The primary scrambling codes have indices $n = 16 \cdot i$ with $i = 0...511$. The ith set of secondary scrambling codes consists of scrambling codes with indices $16 \cdot i + k$ with $k = 1...15$. The primary scrambling code with index i is associated with the ith set of secondary scrambling codes. The 512 primary scrambling codes are further divided into 64 scrambling code groups, each consisting of eight primary scrambling codes. The jth scrambling code group consists of the primary scrambling codes with indices $16 \cdot 8 \cdot j + 16 \cdot l$ with $j = 0...63$ and $l = 0...7$.

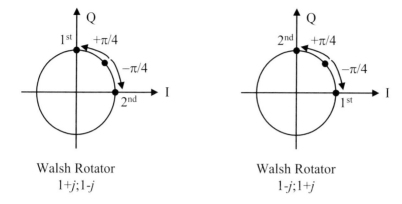

Fig. 6.16. Effect of the decimated second Gold sequence, which randomizes the direction of the $\pi/2$ phase rotation

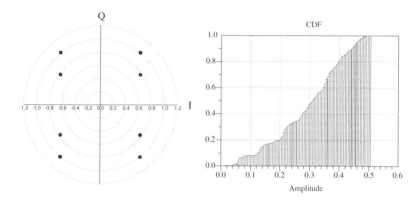

Fig. 6.17. Constellation diagram and CDF for the 768 kbps UL reference measurement channel without scrambling

Only one primary scrambling code is allocated for each cell. The primary scrambling code is always used for transmission of P-CCPCH and P-CPICH. The other DL physical channels are transmitted with either the primary scrambling code or a secondary scrambling code from the set associated with the primary scrambling code of the cell. For the P-CCPCH, the scrambling code is applied aligned with the P-CCPCH frame boundary, i.e., the first complex chip of the spread P-CCPCH frame is multiplied with the first chip of the scrambling code. The scrambling code for the other DL channels is applied aligned with the scrambling code for the P-CCPCH. In this case, the scrambling code is thus not necessarily applied aligned with the frame boundary of the physical channel to be scrambled.

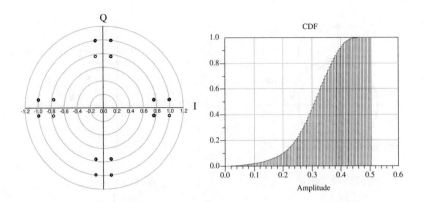

Fig. 6.18. Constellation diagram and CDF for the 768 kbps UL reference measurement channel with scrambling

The complex valued scrambling codes are constructed out of a real Gold code $z_n(i)$, which in turn is generated by a chip-wise modulo 2 sum of two binary m-sequences defined by two generator polynomials of degree 18 (see Appendix C.5.2). The binary sequence $z_n(i)$ with values 0 and 1 is converted to the real valued sequence $Z_n(i)$ with the values $+1$ and -1. Finally, the nth complex scrambling code sequence $S_{\mathrm{dl},n}$ is defined as $S_{\mathrm{dl},n}(i) = Z_n(i) + jZ_n((i + 131,072) \text{ modulo } (2^{18} - 1))$, $i = 0, 1, ..., 38,399$. The pattern from phase 0 up to the phase of 38,399 is repeated.

As opposed to the UL, the scrambling of the DL channels is not designed to yield certain signal properties. It is just a means of distinguishing between different BSs.

6.4.3 Synchronization Codes

The code sequence for the P-SCH is a 256-chip long so-called generalized hierarchical Golay sequence described in Appendix C.6. The 16 SSCs $C_{\mathrm{SSC},k}$, $k = 1, ..., 16$ for the S-SCH are defined with the help of Hadamard codes according to Appendix C.6. Out of the 16 SSCs, 64 different sequences of 15 SSCs are defined in [106], which specify the scrambling code group (section 6.4.2) of a cell. The primary and secondary synchronization channels are not scrambled.

6.4.4 Modulation

The complex valued chip sequence after spreading and scrambling, both in UL and DL is QPSK modulated as shown schematically in Fig. 6.19. The chip rate is 3.84 Mcps. The pulse shaping filter is a root-raised cosine filter according to (2.18) and (2.19) with a roll-off factor of $\alpha = 0.22$ and a chip duration of $T_{\mathrm{chip}} = 0.24042\,\mu\mathrm{s}$. Plots of the impulse response and the magnitude

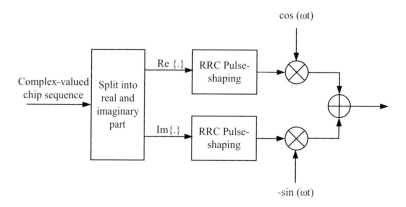

Fig. 6.19. QPSK modulator for up- and downlink

of the frequency transfer function can be seen in Figs. 2.4 and 2.5. Since the pulse shaping is usually implemented by a digital FIR (Finite Impulse Response) filter, values for the number of taps, the number of bits per tap, and the number of samples per chip have to be specified. This non-ideal implementation of the pulse shaping filter introduces a so-called Adjacent Channel Leakage power Ratio (ACLR, see section 6.8.2) and an Error Vector Magnitude (EVM, see section 6.8.2) floor in the system, and even with otherwise perfect components no better ACLR and EVM values can be achieved.

6.5 Multiplexing, Channel Coding, and Interleaving

In this section, a basic overview of the multiplexing, channel coding, and interleaving functions is given. For further details we refer to [109].

The data stream from and to the MAC layer and higher layers is encoded/decoded to offer transport services over the radio transmission link. The channel coding scheme is a combination of error detection, error correction, rate matching, interleaving, and mapping transport channels onto or splitting them from physical channels.

Data arrive at the coding/multiplexing unit in the form of transport block sets once every transmission time interval (TTI). The TTI is transport channel (TrCH) specific and can be 10 ms, 20 ms, 40 ms, or 80 ms. The steps to be performed in the coding and multiplexing scheme are shown in Fig. 6.20 for the UL and in Fig. 6.21 for the DL, respectively. If possible, we will describe the steps for UL and DL together.

CRC. A CRC (Cyclic Redundancy Check) is provided for each transport block. The CRC parity bit block can have a length of 24, 16, 12, 8, or 0 bits.

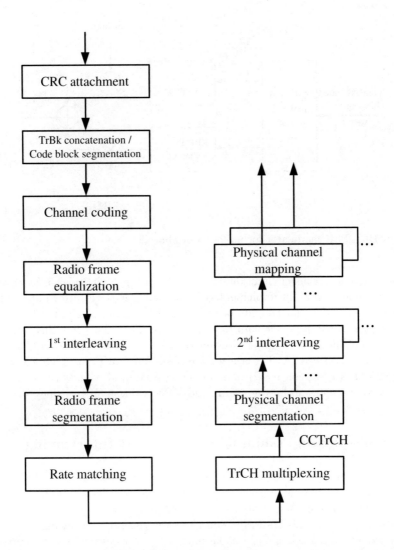

Fig. 6.20. Coding and multiplexing scheme for the uplink

Transport Block Concatenation and Code Block Segmentation. All transport blocks in a TTI are serially concatenated. If the number of bits in a TTI is larger than a certain maximum size of a code block, a code block segmentation is performed after the concatenation of the transport blocks. The maximum size of the code blocks depends on the coding type used for the TrCH and is 504 bits for convolutional coding, 5114 bits for turbo coding, and unlimited in the case of no coding. The code blocks after segmentation

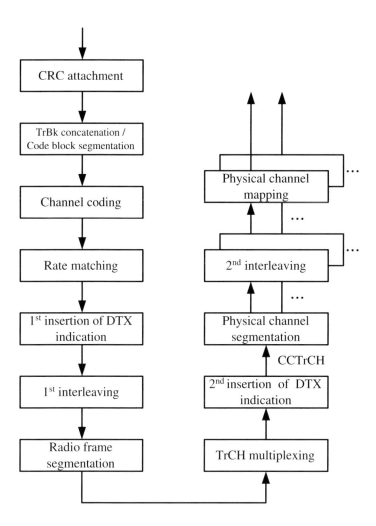

Fig. 6.21. Coding and multiplexing scheme for the downlink

are all of the same size. If necessary, filler bits are added to the beginning of the first block to achieve this. The filler bits are always set to 0.

Channel Coding. The coding schemes and coding rates for the different types of TrCHs are shown in Table 6.6. The convolutional codes with rates 1/2 and 1/3 both have a constraint length of 9. The Turbo coder consists of a Parallel Concatenated Convolutional Code (PCCC) with two 8-state constituent encoders and one Turbo code internal interleaver.

Radio Frame Equalization. Radio frame size equalization is performed in the UL only. The radio frame size equalization ensures that the output bit

Table 6.6. Channel coding schemes and coding rates in FDD mode

Type of TrCH	Coding scheme	Coding rate
BCH	Convolutional coding	1/2
PCH	Convolutional coding	1/2
RACH	Convolutional coding	1/2
CPCH, DCH, DSCH, FACH	Convolutional coding	1/3, 1/2
	Turbo coding	1/3
	No coding	

stream of this functional block can be segmented properly in the radio frame segmentation block. This is done by padding the input bit sequence

Interleaving. Interleaving is performed two times in the UL and DL. The first interleaving operation is placed before the radio frame segmentation if a delay of more than 10 ms due to interleaving is acceptable. Three different interleaving lengths of 20, 40, and 80 ms are defined [75]. The used value depends on the TTI. If in use, the radio frame segmentation block distributes the date from the output of the first interleaver over 2, 4, or 8 consecutive radio frames. The second interleaver operates after the physical channel segmentation and does a 10 ms radio frame (intra-frame) interleaving. The interleaving is applied separately for each physical channel. Both interleavers are block interleavers with inter-column permutations.

Insertion of DTX Indication Bits. In the DL, DTX is used to implement lower data rates than would be possible from the selected SF by turning the transmission off for a short time interval in every slot. The insertion point of DTX indication bits depends on whether fixed or flexible positions of the TrCHs in the radio frame are used. It is up to the UTRAN to decide whether fixed or flexible positions are used during the connection. DTX indication bits only indicate when the transmission should be turned off; they are not transmitted. The first position for the insertion of DTX indication bits is used only if the positions of the TrCHs in the radio frame are fixed. With a fixed position scheme, a fixed number of bits is reserved for each TrCH in the radio frame. Otherwise the DTX indication bits are inserted in the second position. The DTX will be distributed over all slots after the second interleaving.

Radio Frame Segmentation. When the TTI is longer than 10 ms, the input bit sequence is segmented and mapped onto consecutive radio frames. The radio frame equalization block in the UL and the rate matching block in the DL guarantee a proper input bit sequence length for the segmentation.

Rate Matching. The number of bits on a TrCH can vary between different transmission time intervals. To ensure the fixed bit rate assigned for the TrCH, rate matching is performed by means of repetition or puncturing of bits. Puncturing is the omission of bits. Each TrCH is assigned a rate-matching attribute by higher layers. This attribute is semi-static and can only be changed through higher layer signaling. The rate-matching attribute is used to calculate the number of bits to be repeated or punctured.

In the DL, the transmission is interrupted (DTX) if the number of bits is lower than the possible maximum. In the UL, bits are repeated or punctured to ensure that the total bit rate after TrCH multiplexing is identical to the total channel bit rate of the allocated dedicated physical channels. If no bits appear at the input of the UL rate-matching block for all TrCHs, no UL DPDCH will be selected.

Multiplexing of Transport Channels. Every 10 ms, one radio frame from each TrCH is delivered to the TrCH multiplexing. These radio frames are serially multiplexed into a so-called coded composite transport channel (CC-TrCH) which can be mapped to one or several physical channels.

Physical Channel Segmentation. The physical channel segmentation block divides the bits among different physical channels if more than one physical channel is used.

Mapping to Physical Channels. Finally, the bit stream is mapped onto the physical channels, which are described in section 6.3. In the UL, the physical channels during a radio frame are either completely filled with bits to be transmitted or not used at all. The only exception is when the MS operates in the compressed mode (see section 6.7). The transmission can then be turned off during consecutive slots of the radio frame. In the DL the physical channels do not need to be completely filled with bits that are transmitted over the air because of the possibility of DTX.

6.6 Physical Layer Procedures

Some of the most important physical layer procedures are described in this section. Further details can be found in [103].

6.6.1 Cell Search

After a power-on of the MS a full cell search procedure has to be performed. Typically, three steps are carried out during this procedure: slot synchronization, frame synchronization and code group identification, and finally scrambling code identification.

Slot Synchronization. In a first step, slot synchronization is acquired by using the primary synchronization code of the SCH (Fig. 6.5 in section 6.3.8). Since this 256 chip long code is the same in every cell throughout the system and is transmitted at the beginning of every slot, this can easily be performed with a matched filter. By detecting the peaks in the output of the matched filter, the slot timing can be obtained. If more than one peak is detected, as usually will be the case due to multipath propagation and reception from different base stations, the strongest peak will be used, thus synchronizing to the signal from the strongest (nearest) BS.

Frame Synchronization and Code Group Identification. The frame synchronization is found by using the secondary synchronization codes of the SCH. As described in section 6.3.8, 15 SSCs, each 256 chips long, are transmitted over the SCH. In total 16 SSCs and 64 SSC sequences are defined. The 64 sequences are constructed such that their cyclic-shifts are unique, i.e., a non-zero cyclic shift of less than 15 of any of the 64 sequences is not equivalent to some cyclic shift of any other of the 64 sequences. Also, a non-zero cyclic shift of less than 15 of any of the sequences is not equivalent to itself with any other cyclic shift of less than 15. Because of this property the frame boundaries can be found. In addition, the scrambling code group (section 6.4.2) of the cell is encoded in the SSC sequence number.

Scrambling Code Identification. After the detection of the scrambling code group, the exact primary scrambling code used in the cell is determined in the third and last step of the cell search procedure. As each scrambling code group consists of eight primary scrambling codes and the frame boundaries are already known, only eight different correlations have to be performed. These are typically performed over the (primary) common pilot channel which is always scrambled with the primary scrambling code. After the detection of the primary scrambling code, the P-CCPCH, which carries the BCH, can be decoded and the system- and cell specific information can be read.

6.6.2 Power Control

In UMTS, fast closed loop power control with an update rate of 1.5 kHz is applied in DL and UL. Additionally, outer loop power control, which defines the target SIR for the fast closed loop power control, is used. The purpose of power control is mainly to keep the transmit power levels to a minimum. This maximizes the talk times of the MSs and reduces interference, which also optimizes the network capacity. In the UL, the transmitted power causes interference to the adjacent cells, while the received power at the BS causes interference to the received signals of the other users in the same cell. Thus, UL power control minimizes intra- and inter-cell interference. In the DL, the required transmit power limits the network capacity since the DL channels of one user represent interference to all other users.

Under ideal conditions fast power control can compensate for fast fading losses. For a fixed power control update rate this ability of course depends on the speed of the MS. The gain received from power control (either a reduction of the required E_b/N_0 or a reduction of the average transmission power compared to the situation without power control) thus decreases with increasing mobile speed. It has been shown that the use of fast power control for the UMTS FDD mode together with interleaving is able to keep the required E_b/N_0 fairly constant up to mobile speeds of about 250 km/h [110]. At low speed, the power control algorithm is capable of tracking and compensating the changing channel loss. The power control with increasing speed becomes progressively less effective, but on the other hand the interleaver is able to better distribute the error bursts. If the mobile speed increases beyond 250 km/h, significant performance loss occurs and up to $6 - 7$ dB power loss is experienced at a speed of 500 km/h.

If the power control is able to keep the received power level in a fading channel constant, high variations of the transmit power result. This leads to a rise of the average transmit power compared to a non-fading channel. The power rise should be kept to a minimum, since it increases the interference to other cells. It should be noted, that this *rise* in the average transmit power due to fast power control is not a contradiction to the above-mentioned gain from the fast power control in terms of a *reduction* of average transmit power. The reduction of average transmit power is achieved in comparison to the case of no or only slow power control, where a much higher transmit power is necessary in a fast fading environment to guarantee a specified E_b/N_0 than with fast power control. The rise in the average transmit power for a fast fading channel with fast power control is compared to the case of a non-fading channel again with fast power control.

Diversity (e.g., multipath or antenna diversity) is a means to minimize the power rise because diversity enhances the average received power level. Simulation results for the uplink in the UMTS FDD mode showed that with antenna and multipath diversity the average power rise with fast power control decreases with increasing mobile speed [75]. At high speeds the power control is unable to follow the fast fading. Therefore, no power rise is introduced. On the other hand, a higher average received power level is necessary to maintain the link quality. But with high speeds, the diversity effect gained from the multipath diversity is more pronounced, leading to a decrease of the average power rise. An additional rise in the average transmit power is caused by errors in the power control mechanism. The most important sources of errors are the delay in the power control loop, inaccurate SIR estimation and signaling errors.

Most of the power specifications associated with power control in UMTS are based on measurements of the average power in a timeslot with a root-raised cosine filter with a roll-off factor of $\alpha = 0.22$ and a bandwidth equal to the chip rate.

Uplink Closed Loop Power Control. The UL closed loop power control (termed inner-loop power control in the UMTS specification [103]) adjusts the transmit power of the MS in such a way that the received SIR at the BS meets a certain target SIR ($\text{SIR}_{\text{target}}$). If the estimated SIR at the BS is below the target value, the BS issues a power-up command, otherwise a power-down command in transmitted to the MS. The target SIR is determined by the outer loop power control. The transmit powers of the DPCCH and the corresponding DPDCHs (if present) are adjusted by the same amount, if there is no change in the gain factors. The gain factors are determined by the UTRAN. The UL power control operation range is bounded by the minimum output power, which is -50 dBm, and the maximum output power as specified in Table 6.12 in section 6.8.2. The power control commands (named TPC_cmd) are encoded in the TPC bits of the DL DPCCH. Values of $\{-1, 0, +1\}$ are possible for TPC_cmd. The MS computes the change (in dB) of the transmit power of the DPCCH according to

$$\Delta_{\text{DPCCH}} = \Delta_{\text{TPC}} \cdot \text{TPC\_cmd}. \tag{6.12}$$

The step size Δ_{TPC} is specified to be 1, 2, or 3 dB [105], and is controlled by the UTRAN. Table 6.7 shows the necessary accuracy for the different step sizes. The power steps are defined as the relative power difference between

Table 6.7. MS transmit power step size accuracy for the UL closed loop power control

TPC_cmd	1 dB	2 dB	3 dB
±1	±0.5 dB	±1.0 dB	±1.5 dB
0	±0.5 dB		

the average power of the original timeslot and the average power of the target timeslot, not including the transient duration. The transient duration is specified to last from 25 μs before the slot boundary to 25 μs after the slot boundary. If a sequence of 7 or 10 equal TCP commands are transmitted to the base station, the the difference between the original transmit power and the transmit power after the 7 or 10 equal steps must be within tighter limits than 7 or 10 times the limits specified in Table 6.7 (see [105]).

If the MS is in a soft handover procedure, it will receive multiple TCP commands. If these commands are different, the MS combines the commands taking into account the reliability of each TCP command detection to come to a valid decision.

Uplink Open Loop Power Control. Open loop power control is the adjustment of the MS's transmit power to a certain value without any feedback information from the BS. Open loop power control is applied in UTRA FDD only for the random access procedure and the beginning of the CPCH access procedure (see sections 6.6.3 and 6.6.4). The required accuracy of the open loop power control is ±9 dB under normal conditions and ±12 dB under extreme conditions (see Appendix C.1). This large tolerance range reflects the principal uncertainty associated with open loop power control which results from several sources, namely the different propagation conditions in UL and DL, component spread in the MS, and changing temperature.

Downlink Closed Loop Power Control. The operation principles and parameters of the DL power control are essentially the same as for the UL. Concerning the minimum step size for the power control, it is specified that any power changes must be a multiple of the minimum step size $\Delta_{\text{TPC,min}}$. It is mandatory for the BS to support a $\Delta_{\text{TPC,min}}$ of 1 dB and the support of 0.5 dB is optional. The required accuracy for the 1 dB step size is ±0.5 dB, and it is ±0.25 dB for the 0.5 dB step size. The minimum DL power control dynamic range is specified in Table 6.8 [104].

Table 6.8. DL power control dynamic range

Maximum power	BS maximum power −3 dB or higher
Minimum power	BS maximum power −28 dB or lower

Outer Loop Power Control. Outer loop power control adjusts the target SIR for the fast closed loop power control and is applied both in UL and DL. The intention is to maintain a certain required quality of service. If the SIR target is too high, the increased transmission power would give rise to interference, thus wasting capacity. Obviously, a too low SIR target would degrade the link quality. While fast power control has an update rate of 1.5 kHz, outer loop power control operates typically at 10–100 Hz. An issue with outer loop power control is the question what measure to use for monitoring the link quality. One possibility would be to use the result of the CRC check. While this is a simple and reliable method, it is not practicable with high quality services which have frame error rates below 10^{-3}. In this case, a CRC error would occur very rarely, resulting in a much too slow adaptation of the SIR target. Therefore, so-called soft information such as raw BER or soft information from the Viterbi or Turbo decoder will be used in such cases.

6.6.3 Random Access Procedure

Via the RACH (section 6.3.2) and the associated random access procedure tasks like requesting a connection, location updating, or the registration of a MS in the UTRAN are performed. The following steps are carried out during this procedure:

- The MS receives necessary information like the preamble spreading and scrambling codes, timing information, initial open loop power level, power ramping factor, etc., from higher layers.
- The MS randomly selects an allowed RACH sub-channel group and a signature.
- The transmission power is set to the value specified by the RRC layer.
- A preamble 4096 chips long is transmitted to the BS.

- The MS detects the AICH and checks if the BS has received and detected the preamble. If the preamble was not detected (no AICH has been received by the MS), the MS increases the transmit power and transmits the preamble again. This is repeated until either the BS acknowledges successful reception or the maximum number of preamble retransmissions is reached, which leads to an abortion of the random access procedure.
- The MS transmits the message part (10 or 20 ms long) if an acknowledgement of the BS has been received.

6.6.4 CPCH Access Procedure

The CPCH (an UL transport channel, see section 6.3.3) is used for packet data transmission. The structure of the CPCH access procedure is similar to the random access procedure. Through the use of the CPCH Status Indicator Channel the status of the different CPCHs can be monitored by the MS, thus avoiding the attempt to start a CPCH access in the case where all CPCHs are busy. A system option is the use of the channel assignment functionality via the CPCH Collision Detection/Channel Assignment Indicator Channel. This allows the UTRAN to explicitly determine the use of a certain CPCH for the transmission of the message part.

- The first steps of the CPCH access procedure up to either the successful detection of the acquisition indication on the AP-AICH or the abortion of the CPCH procedure are the same as for the random access procedure. The transmitted preamble is called the access preamble (AP).
- If a positive acquisition indicator matched to the chosen preamble signature is detected on the AP-AICH, the access part of the procedure ends and the contention part starts. Now a Collision Detection preamble with length 4096 chips and the same power as the last AP is transmitted.
- The CPCH access procedure is aborted if either no CD/CA-ICH from the BS is received or the channel assignment message tells the MS that all CPCHs are busy. Otherwise the MS transmits an optional power control preamble (with a length of 8 slots) immediately followed by the message part. The physical format of the message part is the same as for the UL DPCH. The length of the message part is an integer multiple of a frame length. The maximum allowed message length is specified by higher layers. Fast closed loop power control is used during the message part.
- If the MS does not detect a start of message indicator on the DL DPCCH in a certain time interval after the start of the transmission, it will terminate the transmission. Otherwise the whole message is transmitted and a success message is issued to the MAC layer.

6.6.5 Downlink Transmit Diversity

The use of DL transmit diversity, i.e., the transmission of the DL signal via two antennas at the BS, is a feature introduced to UMTS to enhance the capacity of the DL. If receive diversity (via two or more antennas) is supported in a BS, the same antennas can also be used for transmit diversity. While not mandatory for the BS, the DL transmit diversity must be supported by the MSs.

UMTS supports two types of transmit diversity techniques, namely open loop and closed loop transmit diversity. With open loop transmit diversity, either space time block coding transmit diversity (STTD) or time switched transmit diversity (TSTD) is used. The following Table 6.9 describes the allowed transmit diversity modes for the DL physical channels. The simul-

Table 6.9. Possible transmit diversity modes for the DL physical channels: X applicable; – not applicable

Physical channel	Open loop		Closed loop
	TSTD	STTD	
P-CCPCH	–	X	–
S-CCPCH	–	X	–
SCH	X	–	–
DPCH	–	X	X
PICH	–	X	–
PDSCH	–	X	X
AICH	–	X	–
CSICH	–	X	–

taneous use of STTD and closed loop modes on the same physical channel is forbidden. If transmit diversity is applied on any of the DL physical channels it must also be applied on the P-CCPCH and the SCH.

Open Loop Transmit Diversity. The open loop transmit diversity is based on STTD for all physical channels except the SCH. A basic block diagram of the STTD coding structure is shown in Fig. 6.22. Channel coding, rate matching and interleaving is performed as in the non-diversity mode. With STTD encoding, blocks of four consecutive channel bits are grouped together and transmitted unchanged via antenna 1. For the second transmission, path order and signs (in part) of the bits are changed.

For the SCH a different type of transmit diversity, so-called TSTD, can be applied. The structure of the transmit diversity scheme is depicted in Fig. 6.23. The codes transmitted on primary and secondary SCH (see Fig. 6.5) are evenly distributed between the two antennas. In the even numbered slots, P-SCH and S-SCH are transmitted via antenna 1, and in odd numbered slots, they are transmitted via antenna 2.

Closed Loop Transmit Diversity. A principal block diagram of the transmitter structure for closed loop mode transmit diversity is shown in Fig. 6.24. The signal processing up to the spreading and scrambling is the same as with-

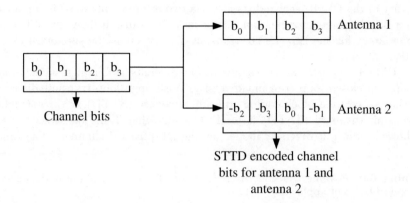

Fig. 6.22. Block diagram of the space time block coding transmit diversity (STTD) scheme

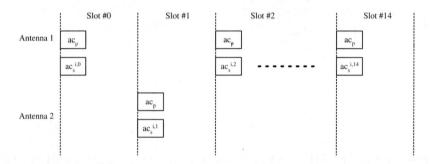

Fig. 6.23. Time switched transmit diversity (TSTD) scheme for the SCH

out transmit diversity. After scrambling, the signal is fed to the two antenna branches, and each signal is multiplied with a distinct and, in general, complex weight factor (w_1 and w_2). The weight factors are controlled by the FBI bit in the UL DPCCH. The CPICHs) are added at both antennas after the weighting. The same channelization and scrambling codes are used for the pilot channels. While the $CPICH_1$ consists of the same fixed symbol sequence as in the case of no transmit diversity, the symbol sequence of the second CPICH ($CPICH_2$) is modified by simply changing the sign of certain symbols [102]. The MS uses the CPICHs to separately estimate the mobile radio channels for both antennas. From the estimated impulse response, the weighting factors are computed once every slot and transmitted to the BS. Two different modes of operation are possible with closed loop transmit diversity.

Closed Loop Mode 1. In closed loop mode 1 only phase adjustments are made by the weighting factors. The phase of the signal transmitted via an-

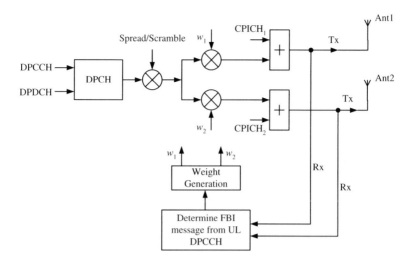

Fig. 6.24. Transmitter structure for closed loop mode transmit diversity

tenna 1 is fixed ($w_1 = 1$) while the phase of the signal transmitted via antenna 2 is adjusted to maximize the received power at the MS. The actual phase adjustment value w_2 is computed in the BS with a sliding window average over the FBI bits in two consecutive slots. Since the FBI field contains one bit this results in four different phase adjustment values. In addition to the two different symbol sequences on the CPICHs, different (orthogonal) dedicated pilot symbols in the DPCCH are also sent over the 2 different antennas.

Closed Loop Mode 2. With closed loop mode 2, phase *and* amplitude weighting is applied. Each feedback command consists of four bits spread over four consecutive slots. One bit controls the amplitude weighting out of the values $\{0.2, 0.8\}$. The remaining three bits allow for eight different phase adjustments in the range from $-180°$ to $+135°$. Progressive updating is performed to achieve an update rate of $1.5\,\text{kHz}$. For closed loop mode 2 the same dedicated pilot symbols in the DPCCH are sent to both antennas.

Performance Gain from Transmit Diversity. For a description of the reasons behind the performance enhancement caused by transmit diversity, we have to distinguish different scenarios. First, we consider the case of a flat fading channel. The two DL transmit signals arrive at the single receive antenna via two different paths and within a chip duration. In the case of closed loop transmit diversity, the relative phase of the two signals are adjusted according to the FBI to achieve maximum received power. We call this coherent combining, which could give a gain of $3\,\text{dB}$ in the ideal case, since the interference is combined non-coherently. In reality, the coarse quantization of the phase adjustment, imperfect channel estimation, and the delay in the feedback loop reduce the gain. The situation changes if we have to deal with a frequency selective mobile radio channel (the multipath components arrive with time differences of more than one chip duration). In this case, each multipath component has a different phase, thus reducing the combining gain further. But the two transmit signals still arrive via two different fading channels, which provides diversity gain not present in the flat fading case.

Unlike the DL, we experience receive diversity in the UL, since the MS's signal arrives via the two different antennas (paths). In this case, the combining is done by a RAKE receiver (section 4.2.2), thus a gain is possible in the range of 2.5 dB up to the maximum value of 3 dB for the ideal case. The actual value depends on the accuracy of the channel estimation [75].

6.6.6 Handover

A handover is the process of switching a MS's connection from one BS to another. This occurs, e.g., if the MS travels from one cell area into the neighboring cell. It is also possible that a handover is issued by the network, e.g., to balance the traffic between neighboring cells. The two basic handover categories are the *hard handover* and the *soft handover*. In the case of a hard handover the MS switches from one BS to another in a way that it is never connected to two BSs simultaneously. In GSM, hard handovers are performed. When the MS communicates with two or more BSs during the handover, this is termed a soft handover, which is only possible if the communication to all BSs takes place in the same frequency band. Soft handovers are performed usually in CDMA-based systems. Other possible handover categories are as follows:

- An intra-mode handover from one BS to another, all of which operate in the UTRA FDD mode. This type of handover can be performed as soft or hard handover. In the case of hard handovers, intra- or inter-frequency handover can be applied. In UMTS, a *softer* handover is possible between different sectors of the same BS.
- The inter-mode handover switches between FDD and TDD mode.
- An inter-system handover takes place between UMTS and another cellular standard.

Further details on handovers in UMTS can be found in [111, 112]. Concerning the physical layer described here, the relevant parameters for handovers are the necessary measurements for determining when and how to perform a handover. These quantities, which are measured by the MS, are the following [113]:

- The Received Signal Code Power (RSCP), which is the received power on one code measured on the Primary CPICH after despreading.
- The Received Signal Strength Indicator (RSSI). This is the wide-band received power within the channel bandwidth.

- E_c/N_0, the received energy per chip divided by the power density in the band. E_c/N_0 is identical to RSCP/RSSI.

Additional measurements have to be performed to obtain the relative timing information between the cells involved in the handover. It is necessary to adjust the frame timing of the new cell in order to be able to allow for coherent combining of the signals of both cells in the RAKE receiver.

If measurements in frequency bands outside the actual receiving band are to be performed, e.g., for an inter-frequency, inter-mode, or inter-system handover, the MS can either use idle periods in the DL transmission generated by using the compressed mode (see section 6.7) or use a second built-in receiver if present.

6.7 Compressed Mode

In the UMTS FDD mode, the so-called compressed mode is foreseen. Transmission and reception are stopped for a short period of time to perform measurements or acquire a control channel from another system or in another frequency band. This measurement/acquisition capability is necessary to support, e.g., a handover. A typical application is for measurements in the GSM 1800 band. Owing to the close proximity of the FDD UL band, simultaneous transmission and reception would be impossible for a MS. Performing measurements in the GSM 900 band while transmitting in the FDD band is technically feasible, but depends on the capability of the MS, because additional active and passive components are required.

As no data should be lost owing to transmission and reception pauses, the data transmission must be compressed before and/or after the idle periods. The possible three different types of compression modes are described in subsection 6.7.2. In most cases compressed frames are used in the DL. If they are necessary in DL and UL they must appear time-aligned with some time offset. The length of the transmission gap (TGL, Transmission Gap Length) is specified to be 3, 4, 7, 10, or 14 slots. The smallest value for the transmission gap length is 3 slots. This is necessary to provide enough time for the hardware to switch to a new frequency, perform the measurement, and switch back again.

During the compressed mode the number of bits assigned to the control data parts on the DPCCH in UL and DL can be different with respect to normal operation. Details can be found in [102]. In the following, we describe some of the specified characteristics of the compressed mode.

6.7.1 Frame Structure

Uplink Frame Structure. The frame structure for UL compressed mode is illustrated in Fig. 6.25.

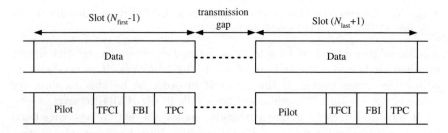

Fig. 6.25. Frame structure for UL compressed mode transmission

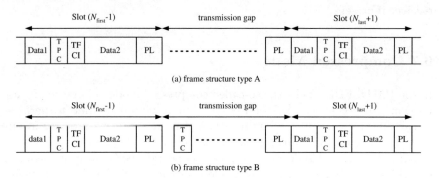

Fig. 6.26. Frame structure for DL compressed mode transmission

Downlink Frame Structure. In the DL two different frame structures are defined for the compressed mode. They are shown in Fig. 6.26. The frame structure A (Fig. 6.26a) maximizes the transmission gap length. With this type only the pilot bits of the last slot in the transmission gap are transmitted. During the rest of the gap the transmitter is turned off. With frame structure B (Fig. 6.26b) the pilot bits of the last slot in the transmission gap and the TPC bits of the first slot in the transmission gap are transmitted and during the rest of the gap the transmitter is turned off. This frame structure is optimized for power control.

6.7.2 Transmission Time Reduction

In the compressed mode the transmission time for the number of bits normally transmitted during a 10 ms frame is reduced. The three mechanisms providing for transmission time reduction are higher layer scheduling, puncturing, and reduction of the spreading factor. All modes are supported in the DL while in the UL puncturing is not used.

Higher Layer Scheduling. The data rate can be lowered by higher layers. They set restrictions in a way that only a subset of the allowed TFCs are

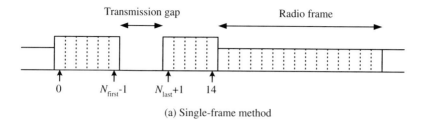

(a) Single-frame method

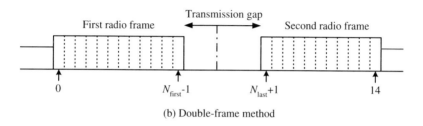

(b) Double-frame method

Fig. 6.27. Transmission gap position

used in compressed mode. Since the maximum number of bits to be delivered to the physical layer during the compressed radio frame is then known, a transmission gap can be generated.

Puncturing. Puncturing can be used to lower the data rate so that a transmission gap can be created. This method is possible only with short transmission gaps owing to limitations of the puncturing. For a detailed description of the puncturing algorithm we refer to [109].

Reduction of the Spreading Factor. In the compressed mode the spreading factor can be reduced by 2 to increase the data rate, which leaves time for the transmission gap.

6.7.3 Transmission Gap Position

The transmission gaps can be placed at different positions. Two different modes are foreseen as is shown in Fig. 6.27. With the single-frame method (Fig. 6.27a) the transmission gap and the compressed data are transmitted in one frame. The transmission gap can be placed anywhere within the radio frame. In the double-frame method (Fig. 6.27b) the transmission gap as well as the compressed slots are divided between two consecutive frames with the transmission gap located at the end of the first and at the beginning of the second frame. Again, the the position of the transmission gap is arbitrary with the only restriction being that at least eight slots must be transmitted in both radio frames. With transmission gap lengths of 3, 4 and 7 slots both types are allowed, while with idle periods of 10 and 14 slots, only the double-frame method can be applied.

6.7.4 Spreading and Scrambling Code Allocation

If the compressed mode is implemented by reducing the spreading factor by 2, the OVSF code used for compressed frames is

- $C_{\text{ch},\text{SF}/2,\lfloor n/2 \rfloor}$ if the "ordinary" scrambling code is used and
- $C_{\text{ch},\text{SF}/2,n \bmod \text{SF}/2}$ if the so-called alternative scrambling code is used.

Here, $C_{\text{ch},\text{SF},n}$ is the channelization code used for the non-compressed frames. To each of the 8192 "ordinary" DL scrambling codes (section 6.4.2) a left and a right alternative scrambling code is associated, which may be used with the compressed frames. The left alternative scrambling code corresponding to scrambling code k has scrambling code number $k + 8192$ while the right alternative scrambling code has scrambling code number $k + 16,384$. If $n < \text{SF}/2$, the left alternative scrambling code is used, otherwise the right alternative scrambling code is applied. Wether an alternative scrambling code on a physical channel should be used or not is signaled by higher layers.

6.7.5 Transmit Power

Owing to the transmission of a higher data rate over a fraction of a frame, the transmit power must be higher during this compressed transmission to achieve the same link quality as in the non-compressed mode. The power step due to the compressed mode must be performed in a way that the energy transmitted on the pilot bits during each transmitted slot follow the inner loop power control. Therefore, the transmit power of the DPCCH in the compressed mode follows the inner loop power control plus additional steps of $\Delta_{\text{PILOT}} = 10 \cdot \log(N_{\text{pilot\_prev}}/N_{\text{pilot\_curr}})$ in dB. Here, $N_{\text{pilot\_prev}}$ is the number of pilot bits in the previously transmitted slot and $N_{\text{pilot\_curr}}$ is the current number of pilot bits per slot. The resulting step of the total transmitted power (DPCCH + DPDCH) is rounded to the closest integer dB value. The accuracy of the power step sizes is given in Table 6.14 in section 6.8.2. The power-versus-time mask that must be met is shown in Fig. 6.28.

Because it is possible that TPC commands for the UL power control are missing during the compressed mode transmission, a so-called recovery period is defined after a transmission gap. During this period some additional features are added to the conventional inner loop power control algorithm. The intention is to recover as rapidly as possible an SIR close to the target SIR after each transmission gap. In the first slot after the transmission gap, a change must be applied in the UL transmit power of the DPCCH with respect to the most recently transmitted slot of $\Delta_{\text{DPCCH}} = \Delta_{\text{RESUME}} + \Delta_{\text{PILOT}}$. Here Δ_{PILOT} is the same as described above. The parameter Δ_{RESUME} is

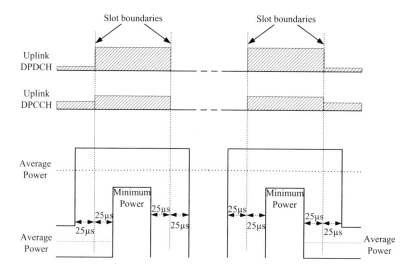

Fig. 6.28. Mask for transmit power change during the compressed mode

determined from parameters received from the UTRAN [103]. The tolerance for this difference in transmission power is specified to be ±3 dB [105]. After the recovery period, ordinary transmit power control is applied again.

6.8 RF Characteristics: Transmitter Requirements, and Receiver Test Cases for the FDD Mode

In this section we will cover the RF characteristics and test cases to be fulfilled by the MSs [105] and BSs [104] and give some general considerations on receiver and transmitter requirements. In general, the transmit and receive characteristics are specified at the antenna connectors of the MS and BS, respectively. Because the antenna performance has an important impact on the system performance, some requirements on the antenna efficiency will probably be included in later releases of the specifications. The power values specified in this section are to be measured after filtering with the same root-raised cosine filter as is used for transmit pulse shaping (section 6.4.4) and by averaging over at least one timeslot excluding any transient periods. Some of the specifications regarding the BS require the measurement of a parameter relative to a value that is not fully specified in the standard documents. In such cases the conformance requirement must be determined relative to a nominal value specified by the manufacturer [114].

The allowed test tolerances for the test and measurement of the specified measurement characteristic are given in [116] and [114] for MS and BS, respectively.

Table 6.10. UARFCN definition

UL	$N_{UL} = 5 \cdot F_{UL}$	$0.0\,\text{MHz} \leq F_{UL} \leq 3276.6\,\text{MHz}$, where F_{UL} is the uplink frequency in MHz
DL	$N_{DL} = 5 \cdot F_{DL}$	$0.0\,\text{MHz} \leq F_{DL} \leq 3276.6\,\text{MHz}$, where F_{DL} is the uplink frequency in MHz

6.8.1 Frequency Bands and Channel Arrangement

As specified in Table 6.1, the so-called paired frequency bands for the UTRA FDD mode are

- $1920 - 1980\,\text{MHz}$ for the UL and
- $2110 - 2170\,\text{MHz}$ for the DL.

In ITU region 2 (which comprises the countries of North- and South-America) these bands are at

- $1850 - 1910\,\text{MHz}$ for the UL and
- $1930 - 1990\,\text{MHz}$ for the DL.

With UTRA FDD both fixed and variable duplex distances are possible. The fixed duplex distance is 190 MHz (80 MHz in region 2) and the variable duplex distance can vary between 134.8 MHz and 245.2 MHz [104].

The nominal channel spacing is 5 MHz. This spacing can be adjusted to optimize the system performance in a particular deployment scenario. The channel center frequency raster to be used for such an adjustment is 200 kHz wide, i.e., the channel center frequency must be an integer multiple of 200 kHz. Within the UTRA framework a carrier frequency is uniquely defined by its UTRA Absolute Radio Frequency Channel Number (UARFCN). The UARFCN is given Table 6.10. From this the valid UARFCN range for the paired band is

- 9612 to 9888 in the UL and
- 10562 to 10838 in the DL.

In ITU region 2 this is

- 9262 to 9538 in the UL and
- 9662 to 9938 in the DL.

6.8.2 MS Transmitter Requirements

The MS transmitter requirements described in the following have to be fulfilled for the 12.2 kbps UL reference measurement channel whose parameters are listed in Table 6.11. The following steps are performed to achieve the DPDCH data rate of 60 kbps from the information bit rate of 12.2 kbps [105]. From the logical channel (see section 6.2.1) DTCH (Dedicated Traffic Channel, a logical channel that is mapped in the MAC layer to a traffic channel),

Table 6.11. Physical parameters for the 12.2 kbps UL reference measurement channel

Information bit rate	12.2 kbps
DPDCH	60 kbps
DPCCH	15 kbps
DPCCH/DPDCH power ratio	-5.46 dB
Power control	On
TFCI	On
Coding	Convolutional coding
Code rate	1/3
Transmission time interval	20 ms

a block consisting of 244 information bits (corresponding to one transmission time interval) is extended with 16 CRC bits and eight tail bits. After convolutional coding with rate 1/3, this block of 268 bits results in 804 bits which is segmented into two radio frames with 402 bits each. After rate matching, we receive 290 bits per radio frame to which 110 bits from the logical channel DCCH (Dedicated Control Channel) are added. A DCCH is a point-to-point bi-directional logical channel designed to carry dedicated control information from higher layers between a MS and the UTRAN. The DCCH must not be confused with the DPCCH which carries only layer 1 control information! This yields 600 bits per radio frame and thus, 60 kbps. The 110 bits per radio frame to be physically transmitted from the DCCH are constructed from 100 DCCH information bits per 40 ms. After adding 12 CRC bits and eight tail bits a convolutional coder with rate 1/3 is applied, yielding 360 bits. These are segmented into four times 90 bits for each radio frame, and after rate matching, we receive 110 bits per frame.

Transmit Power. Four power classes (see Table 6.12) are defined for the MS but only for classes 3 and 4 all RF parameters are specified in the current specifications release. The open loop and inner loop power control step size

Table 6.12. Power classes for MSs

Power class	Maximum output power	Tolerance
1	$+33$ dBm	$+1/-3$ dBm
2	$+27$ dBm	$+1/-3$ dBm
3	$+24$ dBm	$+1/-3$ dBm
4	$+21$ dBm	± 2 dBm

and accuracy have been already described in section 6.6.2. If inner loop and open loop power control indicate minimum transmit output power, the actual transmit power must be at least -50 dBm.

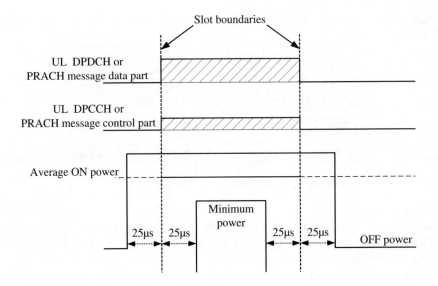

Fig. 6.29. Mask for power ramping between transmit power OFF and ON state

In the transmit OFF power state (the MS transmitter is OFF) the MS's maximum output transmit power within the channel bandwidth must be below -56 dBm. In the case of a random access procedure, packet data transmission (CPCH), or UL compressed mode transitions between transmit ON and OFF state occur. Figure 6.29 shows the time mask for this transitions. The same mask is specified for the PRACH preambles with the only difference being that the length of the ON part is limited to 4096 chips. The accuracy of the transmit power in the ON state depends on the scenario. For the first PRACH preamble, open loop accuracy is sufficient. For transmission after a transmission gap in the compressed mode see section 6.7, and for the maximum transmit power see Table 6.12. During preamble ramping of the RACH/CPCH or between the final RACH/CPCH preamble and the RACH/CPCH message part, the required accuracy depends on the size of the necessary power difference as is shown in Table 6.13.

During UL data transmission, the data rate can vary as indicated by the transmission of the TFCI bits. Usually, an output power change is associated with any change in the TFC (Transport Format Combination). Also, the ratio of the powers between the DPDCH codes and the DPCCH code will vary. Any power change of the DPDCH resulting from a TFC change must follow the inner loop power control. The overall transmit power step (DPCCH + DPDCH) is rounded to the closest integer dB value. Table 6.14 specifies the accuracy of the total transmit power step. The power step is defined as the relative power difference between the average power of the original timeslot and the average power of the target timeslot, not including the transient

Table 6.13. Transmit power step size accuracy for RACH/CPCH preamble ramping and between the final RACH/CPCH preamble and the RACH/CPCH message part

Power step size ΔP in dB	Transmit power step tolerance in dB
0	± 1.0 dBm
1	± 1.0 dBm
2	± 1.5 dBm
3	± 2.0 dBm
$4 \leq \Delta P \leq 10$	± 2.5 dBm
$11 \leq \Delta P \leq 15$	± 3.5 dBm
$16 \leq \Delta P \leq 20$	± 4.5 dBm
$21 \leq \Delta P$	± 6.5 dBm

Table 6.14. Transmit power step size accuracy

Power step size ΔP in dB	Transmit power step tolerance in dB
0	± 0.5 dBm
1	± 0.5 dBm
2	± 1.0 dBm
3	± 1.5 dBm
$4 \leq \Delta P \leq 10$	± 2.0 dBm
$11 \leq \Delta P \leq 15$	± 3.0 dBm
$16 \leq \Delta P \leq 20$	± 4.0 dBm
$21 \leq \Delta P$	± 6.0 dBm

duration. The transient duration is defined from 25 µs before to 25 µs after the slot boundary. The power ramping mask to be fulfilled is shown in Fig. 6.30.

Frequency Accuracy. The carrier frequency accuracy is specified to be within ± 0.1 ppm observed over a period of one timeslot compared to the carrier frequency received from the BS. The same frequency source must be used for RF frequency generation and chip clock.

RF Emissions. The occupied bandwidth, which is defined as containing 99% of the total integrated power of the transmitted spectrum centered on the assigned channel frequency, must be less than 5 MHz for a chip rate of 3.84 Mcps. The spectrum emission masks cover

- the out-of-band emissions, which are between 2.5 MHz and 12.5 MHz away from the carrier frequency, resulting from modulation and nonlinearities in the transmitter (see Fig. 6.31) and

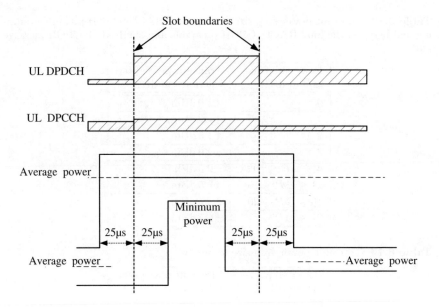

Fig. 6.30. Mask for power ramping due to a change in the transport format combination

- the spurious emissions caused by harmonics emission, parasitic emission, intermodulation products and frequency conversion products (see Fig. 6.32).

The out-of-band emissions are specified relative to the channel output power, which has to be measured with a 3.84 MHz bandwidth. According to the specifications, a measurement bandwidth of 30 kHz has to be used for frequencies up to 3.5 MHz away from the carrier and above a measurement bandwidth of 1 MHz is specified. Therefore, the spectrum emission mask values shown in Fig. 6.31 have to be changed accordingly. For the spurious emissions, different measurement bandwidths, depending on the actual frequency band, are given in Fig. 6.32, e.g., for the GSM bands more stringent limits apply. The spurious emission mask only applies to frequencies more than 12.5 MHz away from the carrier frequency.

Besides the emission masks, a second measure for emission limits is the *Adjacent Channel Leakage power Ratio* (ACLR). The ACLR is defined as the ratio of the transmitted power to the power measured in an adjacent channel. Both, the transmitted power and the adjacent channel power are measured with a root-raised cosine filter with a roll-off factor of $\alpha = 0.22$ and a bandwidth equal to the chip rate. The minimum requirement for the ACLR is 33 dB (or −50 dBm absolute adjacent channel power, whichever is higher) for the neighboring channels spaced ±5 MHz away from the transmit channel and 43 dB (or −50 dBm absolute adjacent channel power, whichever

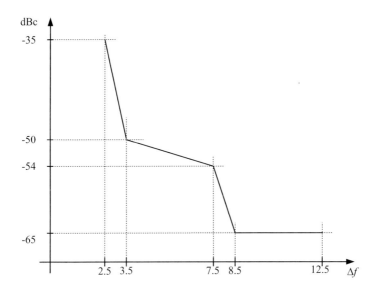

Fig. 6.31. Spectrum emission mask for the out-of-band emissions measured with a bandwidth of 30 kHz

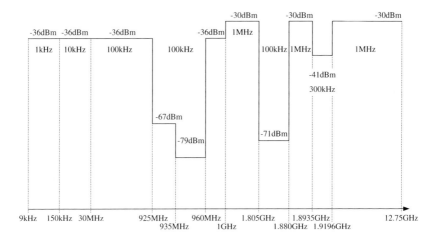

Fig. 6.32. Spectrum emission mask for spurious emissions. For each frequency band the measurement bandwidth is given separately

is higher) for the channels spaced ±10 MHz away. These values are specified for power classes 3 and 4 and must still be met in the presence of switching transients.

Transmit Intermodulation. The transmit intermodulation performance describes the capability of the transmitter to suppress the generation of spurious signals in its nonlinear elements caused by the presence of the wanted signal and an interfering signal that is received via the antenna. Table 6.15 specifies the requirements which must be met if the MS transmits with maximum output power. The MS intermodulation requirement is defined as the

Table 6.15. Transmit intermodulation requirements

Interference signal frequency offset	±5 MHz	±10 MHz
Interference CW signal level	−40 dBc	
Intermodulation product	−31 dBc	−41 dBc

ratio of the intermodulation product caused by an interfering CW signal with a level below the wanted signal to the output power of the wanted signal. Both the wanted signal power and the intermodulation product power are measured with a root-raised cosine filter with a roll-off factor of $\alpha = 0.22$ and a bandwidth equal to the chip rate. Owing to amplification in the transmitter, the intermodulation product can be higher than the interfering signal, which is reflected in the specification.

Transmit Modulation. The pulse shaping filter in the transmitter is a root-raised cosine filter (see (2.18)) with a roll-off factor of $\alpha = 0.22$ and a bandwidth of 3.84 MHz which corresponds to the chip duration of $T_{\text{chip}} = 0.26042\,\mu s$. The accuracy of the vector modulation is specified in terms of the *Error Vector Magnitude (EVM)*. EVM is a measure of the difference between the measured waveform and the theoretical modulated waveform. It is defined as the square root of the ratio of the mean error vector power to the mean reference signal power expressed in percent. The measurement interval is one power control group (timeslot). The EVM of the whole transmitter must stay below 17.5% if the MS's output power is above −20 dBm and the power control step size is 1 dB.

According to Annex B in [116], the EVM is calculated as follows. We define **Z** as a vector containing the samples of the output signal of the transmitter under test after root-raised cosine ($\alpha = 0.22$) filtering. **Z** contains the signal of the whole measurement interval with one sample per chip. A reference signal vector **R** is generated in the same way as **Z** with the exception that no transmitter under test is present. The reference signal **R** is now varied with respect to delay, phase, amplitude, frequency, DC offset, etc., in order to achieve the best fit with the signal **Z**. This variation is necessary to account for the effects introduced by the transmitter under test, e.g., filtering, amplification, mixing. Best fit is achieved when the RMS (root mean

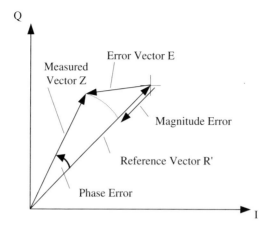

Fig. 6.33. Definition of the error vector

square) difference between **Z** and the varied reference signal is an absolute minimum. The varied reference signal after the best fit process is called **R′**. The difference between **R′** and **Z** is the error vector **E**

$$\mathbf{E} = \mathbf{Z} - \mathbf{R}'. \tag{6.13}$$

Figure 6.33 illustrates this equation for an example chip value. The following steps have to be performed to compute the EVM.

1. Calculation of RMS(**E**), the RMS value of the error vector **E**.
2. Calculation of RMS(**R′**), the RMS value of the varied reference signal **R′**.
3. Calculation of the EVM (in %) according to

$$\mathrm{EVM} = \frac{\mathrm{RMS}(\mathbf{E})}{\mathrm{RMS}(\mathbf{R}')} \cdot 100. \tag{6.14}$$

Further details on the practical implementation of the EVM measurement can, e.g., be found in [117].

The *Peak Code Domain Error (PCDE)* is computed by projecting the power of the error vector onto the code domain at a specific spreading factor. The Code Domain Error for every code in the domain is defined as the ratio of the mean power of the projection onto that code to the mean power of the composite reference waveform expressed in dB. The PCDE is now defined as the maximum value for the Code Domain Error for all codes. The measurement interval is one timeslot. The definition and measurement of the PCDE makes sense only for multi-code transmission. The PCDE must stay below −15 dB at a SF of 4 for the same parameters specified for the EVM measurement but using the 768 kbps UL reference measurement channel whose characteristic is listed in Table 6.16. For the computation of the PCDE the

Table 6.16. Physical parameters for the 768 kbps UL reference measurement channel

Information bit rate	$2 \cdot 384\,\text{kbps} = 768\,\text{kbps}$
DPDCH$_1$	960 kbps
DPDCH$_2$	960 kbps
DPCCH	15 kbps
DPCCH/DPDCH power ratio	$-11.48\,\text{dB}$
TFCI	On
Coding	DTCH: convolutional coding DCCH: Turbo coding
Code rate	1/3

error vector $\mathbf{E}$, as is used for the computation of the EVM, is decomposed into time-sequential vectors $\mathbf{e}$, each with SF complex samples comprising one symbol interval. Since SF $= 4$, each vector $\mathbf{e}$ consists of 4 samples. The following steps have to be performed to compute the PCDE.

1. Computation of $\mathbf{e}'$ by descrambling the vectors $\mathbf{e}$.
2. Calculation of the inner product of $\mathbf{e}'$ with all k codes of the channelization codes set $\mathbf{C}$ for all ns symbols of the measurement interval. This gives a $k \times ns$ array with each error value defined by a specific symbol and a specific code.
3. Calculation of k RMS values, each RMS value unifying ns symbols within one code. These values are called "Absolute CodeEVMs".
4. Find "Absolute PeakCodeEVM", the peak value among the k "Absolute CodeEVMs".
5. Calculation of the PCDE (in dB) according to:

$$\text{PCDE} = 10 \cdot \log \frac{(\text{Absolute PeakCodeEVM})^2}{(\text{RMS}(\mathbf{R}'))^2} . \tag{6.15}$$

6.8.3 BS Transmitter Requirements

The following test conditions have to be applied for the BS transmitter specifications described subsequently except where otherwise noted. To create a realistic traffic scenario with a high peak-to-average power ratio a total of 64 DPCHs with 30 ksps (SF $= 128$, slot format number 10, see Appendix C.2) distributed randomly across the code space are defined at random power levels and random timing offsets. The DPDCH and DPCCH of one DPCH have the same power level. Test scenarios with 32 and 16 DPCHs are also specified because not every base station will support 64 DPCHs. The conformance tests have to be performed using the test scenario with the largest possible number of DPCHs. Table 6.17 lists the active channels for the test scenario. The column "Fraction of power" specifies the power of the channel relative

Table 6.17. Active channels for the BS test scenario

Type	Number	Fraction of power (%)	Level setting (dB)	Channel. code	Timing offset ($\times 256 \cdot T_{\text{chip}}$)
PCCPCH + SCH	1	10	-10	1	0
P-CPICH	1	10	-10	0	0
PICH	1	1.6	-18	16	120
S-CCPCH	1	1.6	-18	3	150
DPCH	16/32/64	76.6 in total	[1]	[1]	[1]

[1] For detailed parameters of the DPCHs, see [114].

Table 6.18. Spectrum emission mask values for a BS maximum output power $P \geq 43\,\text{dBm}$

Δf	f_{off}	Maximum emission level	Measurement bandwidth
MHz	MHz	dBm	kHz
$2.5 \leq \Delta f < 2.7$	$2.515 \leq f_{\text{off}} < 2.715$	-14	30
$2.7 \leq \Delta f < 3.5$	$2.715 \leq f_{\text{off}} < 3.515$	$-14 - 15(f_{\text{off}} - 2.715)$	30
	$3.515 \leq f_{\text{off}} < 4.0$	-26	30
$3.5 \leq \Delta f$	$4.0 \leq f_{\text{off}} < f_{\text{off,Max}}$	-13	1000

to the maximum output power at the transmitter antenna interface under test. Further details on the channels used (channelization code, power level, and timing offset) are given in [114].

Frequency Accuracy. The carrier frequency accuracy is specified to be within $\pm 0.05\,\text{ppm}$ observed over one timeslot. The same frequency source must be used for RF frequency generation and chip clock.

Out-of-Band Emissions. The spectrum emission masks given in Tables 6.18, 6.19, 6.20, and 6.21 and summarized in Fig. 6.34 cover out-of-band emissions resulting from modulation and nonlinearities in the transmitter. The parameter Δf is the frequency separation between the carrier frequency and the $-3\,\text{dB}$ point of the measuring filter closest to the carrier frequency. The separation between the carrier frequency and the center frequency of the measurement filter is denoted by f_{off} and $f_{\text{off,Max}}$ is either $12.5\,\text{MHz}$ or the offset to the UMTS transmit band edge, whichever is higher.

As a second measure for emission limits the ACLR (defined in section 6.8.2) must be better than $45\,\text{dB}$ for the neighboring channels spaced $\pm 5\,\text{MHz}$ away from the transmit channel and $50\,\text{dB}$ for the channels spaced $\pm 10\,\text{MHz}$ away.

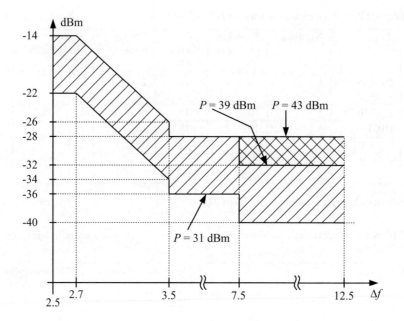

Fig. 6.34. Spectrum emission mask for the out-of-band emissions measured with a bandwidth of 30 kHz

Table 6.19. Spectrum emission mask values for a BS maximum output power $39 \leq P < 43$ dBm

Δf	f_{off}	Maximum emission level	Measurement bandwidth
MHz	MHz	dBm	kHz
$2.5 \leq \Delta f < 2.7$	$2.515 \leq f_{\mathrm{off}} < 2.715$	-14	30
$2.7 \leq \Delta f < 3.5$	$2.715 \leq f_{\mathrm{off}} < 3.515$	$-14-15(f_{\mathrm{off}}-2.715)$	30
	$3.515 \leq f_{\mathrm{off}} < 4.0$	-26	30
$3.5 \leq \Delta f < 7.5$	$4.0 \leq f_{\mathrm{off}} < 8.0$	-13	1000
$7.5 \leq \Delta f$	$8.0 \leq f_{\mathrm{off}} < f_{\mathrm{off,Max}}$	$P - 56$	1000

Spurious Emissions. The spurious emission limits for BS transmitters are divided into category A and B limits according to the ITU recommendation SM.329-8 [115]. In terms of the ITU, category A specifies general limits, while category B describes more stringent limits, especially based on limits defined in Europe. For category A, the spurious emission limits are listed in Table 6.22 and for category B in Table 6.23. The frequencies F_{c1} and F_{c2} are the center frequencies of the first and last carrier transmitted by the BS, respectively. To protect the highly sensitive BSs receiver, the spurious emissions in the receive band (1920 − 1980 MHz or 1850 − 1910 MHz) must

Table 6.20. Spectrum emission mask values for a BS maximum output power $31 \leq P < 39\,\mathrm{dBm}$

Δf	f_off	Maximum emission level	Measurement bandwidth
MHz	MHz	dBm	kHz
$2.5 \leq \Delta f < 2.7$	$2.515 \leq f_\mathrm{off} < 2.715$	$P - 53$	30
$2.7 \leq \Delta f < 3.5$	$2.715 \leq f_\mathrm{off} < 3.515$	$P - 53 - 15(f_\mathrm{off} - 2.715)$	30
	$3.515 \leq f_\mathrm{off} < 4.0$	$P - 65$	30
$3.5 \leq \Delta f < 7.5$	$4.0 \leq f_\mathrm{off} < 8.0$	$P - 52$	1000
$7.5 \leq \Delta f$	$8.0 \leq f_\mathrm{off} < f_\mathrm{off,Max}$	$P - 56$	1000

Table 6.21. Spectrum emission mask values for a BS maximum output power $P < 31\,\mathrm{dBm}$

Δf	f_off	Maximum emission level	Measurement bandwidth
MHz	MHz	dBm	kHz
$2.5 \leq \Delta f < 2.7$	$2.515 \leq f_\mathrm{off} < 2.715$	-22	30
$2.7 \leq \Delta f < 3.5$	$2.715 \leq f_\mathrm{off} < 3.515$	$-22 - 15(f_\mathrm{off} - 2.715)$	30
	$3.515 \leq f_\mathrm{off} < 4.0$	-34	30
$3.5 \leq \Delta f < 7.5$	$4.0 \leq f_\mathrm{off} < 8.0$	-21	1000
$7.5 \leq \Delta f$	$8.0 \leq f_\mathrm{off} < f_\mathrm{off,Max}$	-25	1000

Table 6.22. Spurious emission limits for BSs according to category A

Band	Maximum level	Measurement bandwidth
$9\,\mathrm{kHz} - 150\,\mathrm{kHz}$	$-13\,\mathrm{dBm}$	$1\,\mathrm{kHz}$
$150\,\mathrm{kHz} - 30\,\mathrm{MHz}$	$-13\,\mathrm{dBm}$	$10\,\mathrm{kHz}$
$30\,\mathrm{MHz} - 1\,\mathrm{GHz}$	$-13\,\mathrm{dBm}$	$100\,\mathrm{kHz}$
$1\,\mathrm{GHz} - 12.75\,\mathrm{GHz}$	$-13\,\mathrm{dBm}$	$1\,\mathrm{MHz}$

stay below $-96\,\mathrm{dBm}$ for a measurement bandwidth of $100\,\mathrm{kHz}$. The spurious emission limits for the coexistence with other mobile communication standards are given in Table 6.24. For the coexistence with services in frequency bands adjacent to either $2110-2170\,\mathrm{MHz}$ or $1930-1990\,\mathrm{MHz}$ (for ITU region 2), additional spurious emission limits are specified which are listed in Table 6.25. These requirements must be applied in geographic areas in which both an adjacent band service and UTRA are deployed.

Transmit Intermodulation. The power of the intermodulation products at the antenna connector should not exceed the out-of-band emission or the spurious emission requirements listed in the preceding subsections if a W-CDMA modulated interference signal is injected into the antenna connector

Table 6.23. Spurious emission limits for BSs according to category B

Band	Maximum level	Measurement bandwidth
9 kHz − 150 kHz	−36 dBm	1 kHz
150 kHz − 30 MHz	−36 dBm	10 kHz
30 MHz − 1 GHz	−36 dBm	100 kHz
1 GHz − (F_{c1} − 60 MHz) or 2.1 GHz whichever is higher	−30 dBm	1 MHz
(F_{c1} − 60 MHz) or 2.1 GHz whichever is higher − (F_{c1} − 50 MHz) or 2.1 GHz whichever is higher	−25 dBm	1 MHz
(F_{c1} − 50 MHz) or 2.1 GHz whichever is higher − (F_{c2} + 50 MHz) or 2.18 GHz whichever is lower	−15 dBm	1 MHz
(F_{c2} + 50 MHz) or 2.18 GHz whichever is lower − (F_{c2} + 60 MHz) or 2.18 GHz whichever is lower	−25 dBm	1 MHz
(F_{c2} + 60 MHz) or 2.18 GHz whichever is lower − 12.75 GHz	−30 dBm	1 MHz

Table 6.24. BS spurious emission limits for the coexistence with other mobile communication standards

Band	Maximum level	Measurement bandwidth	Note
876 − 915 MHz	−98 dBm	100 kHz	GSM 900 BTS receiver
921 − 960 MHz	−57 dBm	100 kHz	GSM 900 MS receiver
1710 − 1785 MHz	−98 dBm	100 kHz	co-location with DCS 1800 BTS receiver
1805 − 1880 MHz	−47 dBm	100 kHz	DCS 1800 MS receiver
1893.5 − 1919.6 MHz	−41 dBm	100 kHz	PHS
1900 − 1920 MHz	−52 dBm	1 MHz	if UTRA TDD BSs are located in the same geographic area
2010 − 2025 MHz	−52 dBm	1 MHz	if UTRA TDD BSs are located in the same geographic area
1900 − 1920 MHz	−86 dBm	1 MHz	for colocated UTRA TDD BSs
2010 − 2025 MHz	−86 dBm	1 MHz	for colocated UTRA TDD BSs

at a level of 30 dB below that of the transmit signal. The frequency offsets of the interfering signal are ±5 MHz, ±10 MHz, and ±15 MHz from the transmit signal.

Table 6.25. Spurious emission limits for the coexistence with services in adjacent frequency bands

Band	Maximum level	Measurement bandwidth
$2100 - 2105\,\mathrm{MHz}$ [1]	$-30 + 3.4(f - 2100\,\mathrm{MHz})\,\mathrm{dBm}$	1 MHz
$2175 - 2180\,\mathrm{MHz}$ [1]	$-30 + 3.4(2180\,\mathrm{MHz} - f)\,\mathrm{dBm}$	1 MHz
$1920 - 1925\,\mathrm{MHz}$ [2]	$-30 + 3.4(f - 1920\,\mathrm{MHz})\,\mathrm{dBm}$	1 MHz
$1995 - 2000\,\mathrm{MHz}$ [2]	$-30 + 3.4(2000\,\mathrm{MHz} - f)\,\mathrm{dBm}$	1 MHz

[1] If operated in the frequency bands $1920 - 1980\,\mathrm{MHz}$ (UL) and $2110 - 2170\,\mathrm{MHz}$ (DL).

[2] If operated in the frequency bands $1850 - 1910\,\mathrm{MHz}$ (UL) and $1930 - 1990\,\mathrm{MHz}$ (DL) in ITU Region 2 which comprises the countries of North and South America.

Transmit Modulation. For the transmit modulation accuracy characterization the same parameters (EVM and PCDE) with the same definitions as for the MS are used. For the EVM also the limit of 17.5 % is the same as for the MS. The PCDE for the BS must not exceed $-33\,\mathrm{dB}$ at a spreading factor of 256. The test signals for EVM and PCDE are different from the test signal used for all other BS transmitter requirements. For the EVM measurement, besides the DPCH, the signal consists of the PCCPCH and the SCH with channelization code 1 and timing offset 0, which together make up for 1.6 to 50 % of the total transmit power (level setting of -18 to $-3\,\mathrm{dB}$). An optional P-CPICH with channelization code 0, timing offset 0, and 10% of the total power can be included. The test signal composition for the PCDE measurements are given in Table 6.26. If a BS does not support 32 channels the parameters for 16 channels have to be used.

Table 6.26. Active channels for the BS test signal for the PCDE measurements

Type	Number	Fraction of power (%) 16/32	Level setting (dB) 16/32	Channel. code	Timing offset ($\times 256 \cdot T_{\mathrm{chip}}$)
P-CCPCH + SCH	1	12.6/7.9	$-9/-11$	1	0
P-CPICH	1	12.6/7.9	$-9/-11$	0	0
PICH	1	5/1.6	$-13/-18$	16	120
S-CCPCH containing PCH	1	5/1.6	$-13/-18$	3	150
DPCH	16/32	63.7/80.4 in total	[1]	[1]	[1]

[1] For detailed parameters of the DPCHs, see [114].

6.8.4 MS Receiver Test Cases

Table 6.27 lists common terms used in the UMTS specifications [105, 129] for defining the MS receiver requirements to be fulfilled. Unless otherwise stated, all parameters are specified at the antenna connector of the MS. They are defined using the 12.2 kbps DL reference measurement channel with

Table 6.27. Abbreviations used for the description of the UTRA FDD RF characteristics

$DPCH\_E_C$	Average energy per chip of a DPCH
$\hat{I}_{or}$	Received (DL) power spectral density measured at the MS antenna connector
I_{or}	Total DL transmit power spectral density at the base station antenna connector
I_{oac}	Power spectral density of the adjacent channel measured at the MS antenna connector
I_{ouw}	Power level (or power spectral density in case of a modulated signal) of the unwanted signal measured at the MS antenna connector
OCNS	Orthogonal Channel Noise Simulator , a mechanism used to simulate users or control signals on the other orthogonal channels of a downlink

the specifications listed in Table 6.28. This describes a downlink dedicated

Table 6.28. Physical parameters for the 12.2 kbps DL reference measurement channel

Information bit rate	12.2 kbps
DPCH	30 ksps
Slot format #	11
Power offsets PO1, PO2, and PO3	0 dB
Puncturing	14.7%
TFCI	On
Power control	Off
Coding	Convolutional coding
Code rate	1/3

physical channel with a spreading factor of 128 (referred to symbol spreading). All terminals must support this type of low data rate channel. To perform the receiver test case measurements under realistic conditions, the common DL physical channels required for a connection between BS and MS must be added to the wanted DPCH. Table 6.29 lists these DL channels, together with

Table 6.29. DL physical channels transmitted in addition to the DPCH during a connection

Physical channel	Power spectral density relative to DPCH
CPICH	7 dB
P-CCPCH	5 dB
SCH	5 dB
PICH	2 dB
DPCH	–

Table 6.30. Reference sensitivity test case

DPCH_E_C	$-117\,\text{dBm}/3.84\,\text{MHz}$
$\hat{I}_{or}$	$-106.7\,\text{dBm}/3.84\,\text{MHz}$

their relative power spectral densities with respect to the DPCH [116], which resemble the reference measurement setup. If the powers of all channels are added, taking into account that P-CCPCH (which carries the BCH) and SCH are time multiplexed, the ratio of the total transmitted power to the transmit power of the DPCH amounts to 10.3 dB. This channel configuration must be applied to all receiver test cases except that for the maximum input level. If not otherwise stated, the coded bit error rate under the conditions described for each receiver test case must not exceed 10^{-3}.

Reference Sensitivity Test Case. The reference sensitivity test case specifications are given in Table 6.30. The requirement of BER $< 10^{-3}$ must be fulfilled during simultaneous operation of the MS's transmitter at maximum output power. From this test case a simple calculation yields the required noise figure of the whole receiver. The noise power in the UMTS channel bandwidth of $B = 3.84\,\text{MHz}$ is given as

$$P_N = 10\log(kTB) = -108\,\text{dBm} \tag{6.16}$$

with the Boltzmann constant $k = 1.38 \cdot 10^{-23}\,\text{W/KHz}$, and the noise temperature $T = 300\,\text{K}$. From simulations [118], a necessary E_b/N_t was determined to be 5.2 dB to achieve a BER of 10^{-3}. Here E_b/N_t is the ratio of energy per information bit to the total effective noise and interference power spectral density, including thermal noise and all other interference. If we account for an implementation loss in the baseband signal processing of $1 - 2\,\text{dB}$, we obtain an effective necessary value of $(E_b/N_t)_{\text{eff}} = 7.2\,\text{dB}$. In [110] values of 7.5–8.5 dB have been reported, but in this case multipath channels have been implemented in the simulations. Since the reference sensitivity test case is defined for an AWGN channel, it can be concluded that the stated $(E_b/N_t)_{\text{eff}}$ of 7.2 dB is a reasonable assumption. The processing gain caused by spreading (SF $= 128$) is

Table 6.31. Maximum input level test case

$\frac{\text{DPCH\_E}_{\text{C}}}{I_{\text{or}}}$	$-19\,\text{dB}$
$\hat{I}_{\text{or}}$	$-25\,\text{dBm}/3.84\,\text{MHz}$

Table 6.32. DL physical channels which are active in addition to the DPCH during the measurement for the maximum input level test case

Physical channel	Power spectral density relative to DPCH
P-CPICH	10 dB
S-CPICH	10 dB
P-CCPCH	10 dB
SCH	12 dB
PICH	15 dB
OCNS	Necessary power so that I_{or} adds to one

$$G_{\text{p}} = 10\log\left(\frac{\text{Chip Rate}}{\text{Symbol Rate}}\right) = 10\log\left(\frac{3.84\,\text{Mcps}}{30\,\text{ksps}}\right)$$
$$= 10\log(128) = 21\,\text{dB}. \qquad (6.17)$$

For the gain due to convolutional coding an estimation is difficult. We assume a coding gain G_{c} of 4 dB which is rather conservative. The required NF of the whole receiver chain can now be computed according to

$$\text{NF} = P_{\text{DPCH}} + G_{\text{p}} + G_{\text{c}} - \left(\frac{E_{\text{b}}}{N_{\text{t}}}\right)_{\text{eff}} - P_{\text{N}}$$
$$= -117\,\text{dBm} + 21\,\text{dB} + 4\,\text{dB} - 7.2\,\text{dB} + 108\,\text{dBm}$$
$$= 8.8\,\text{dB}. \qquad (6.18)$$

Maximum Input Level Test Case. The maximum input level test case specifies the maximum received input power at the MS antenna connector, which should not degrade the specified BER performance of 10^{-3}. Table 6.31 specifies the parameters. For power classes 3 and 4, the average transmit output power must be 20 dBm and 18 dBm, respectively. Table 6.32 lists the DL physical channels which are active the measurement of the maximum input level test case. For this test case the difference between the total received power spectral density in the receive band and that of the DPCH is -19 dB, which is much larger than for the other test cases (as described at the beginning of this section). A part of this signal power comes from additional users on the same channel. In the UMTS specifications this type of interference is modeled with an OCNS (Orthogonal Channel Noise Simulator) signal. OCNS is defined as a mechanism to simulate users or control signals on the other orthogonal channels of the DL. The OCNS signal consists of 16 dedicated data channels with data uncorrelated both to each other and the reference

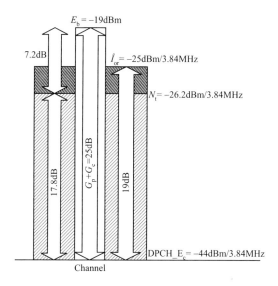

Fig. 6.35. Maximum input level test case

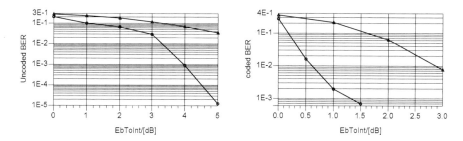

Fig. 6.36. Uncoded and coded bit error rate (BER) for interference modeled as Gaussian noise (▲) and OCNS signal (●)

measurement channel. In Appendix C.7, the DPCH spreading codes, timing offsets, and relative levels for the 16 DPCH's are listed.

A graphical representation of this test case is shown in Fig. 6.35. If we make the same assumptions as in the reference sensitivity test case (25 dB combined processing, and coding gain and 7.2 dB required effective signal to interference ratio) the allowed interference power spectral density computes to $N_t = -26.2$ dBm/3.84 MHz. For this test case, however, these simplistic assumptions are not valid. The simulation results shown in Fig. 6.36 indicate that E_b/N_t can be as low as 1.5 dB to achieve a BER of 10^{-3}. The majority of the interference power in this test case comes from other users. Because the signals from other users are orthogonal, they are greatly suppressed by the despreading operation in the receiver, resulting in the low required E_b/N_t.

The assumed combined processing and coding gain of 25 dB is valid only for an orthogonality factor of $OF = 0$ dB (see sections 3.2.1 and 4.2.1). For the maximum input signal test case the interference coming from other users has a higher OF. This is also illustrated in Fig. 6.36 where BER simulation results for interference modeled as Gaussian noise ($OF = 0$ dB) and as OCNS signal ($OF > 0$ dB) are compared.

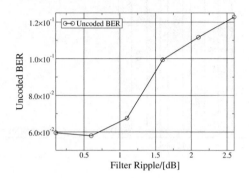

Fig. 6.37. Uncoded BER as function of the in-band peak-to-peak amplitude ripple of a Chebyshev-type band-pass filter

Owing to the high power level of the received signal, thermal noise does not need to be considered. Under ideal conditions the OCNS signal would be completely suppressed by the despreading in the receiver. Therefore, this test case implicitly defines the allowable in-band distortions introduced in the receiver. A major source of such in-band distortions is the channel selection filtering at the baseband or at the IF (intermediated frequency) in the case of a heterodyne receiver. To illustrate the effects of filtering, Fig. 6.37 [120] displays the uncoded BER as a function of the in-band peak-to-peak amplitude ripple of a Chebyshev-type band-pass filter (implemented as equivalent low-pass FIR filter in the simulation). The distortions can be attributed to amplitude and phase distortion, since no group delay equalization was performed in this simulation. A ripple of 0.6 dB has no visible influence on the BER and up to 1.1 dB only minor degradation is introduced. For higher ripples the in-band distortion results in a noticeable BER degradation. The uncoded BER was chosen as a performance measure in this example, because of simulation time considerations. A further source of in-band distortions, which degrades the orthogonality between different users, and therefore the receiver performance, is the MS receiver front-end nonlinearity (see, e.g., [121]).

Adjacent Channel Selectivity Test Case. The adjacent channel selectivity (ACS) test case determines the ability of the UMTS receiver to suppress a signal in the neighboring channel (spaced 5 MHz away from the wanted channel) with a much higher power than the signal in the wanted channel

Table 6.33. Parameters for the adjacent channel selectivity test case

ACS	33 dB
DPCH_E_C	$-103\,\text{dBm}/3.84\,\text{MHz}$
$\hat{I}_{or}$	$-92.7\,\text{dBm}/3.84\,\text{MHz}$
I_{oac} (modulated)	$-52\,\text{dBm}/3.84\,\text{MHz}$

Table 6.34. Parameters for the in-band blocking test case

DPCH_E_C	$-114\,\text{dBm}/3.84\,\text{MHz}$
$\hat{I}_{or}$	$-103.7\,\text{dBm}/3.84\,\text{MHz}$
$I_{blocking}$ (modulated)	$-56\,\text{dBm}/3.84\,\text{MHz}$ at $\pm 10\,\text{MHz}$ away
$I_{blocking}$ (modulated)	$-44\,\text{dBm}/3.84\,\text{MHz}$ at $\pm 15\,\text{MHz}$ away

without degrading the BER. The situation modeled with the ACS test case occurs if a MS is located at the boundary of its serving cell, which belongs to the network operator to which the user is subscribed, and at the same time a BS of another operator is transmitting nearby in the adjacent channel. The parameters for this test case, specified for the UTRA FDD mode, are listed in Table 6.33. For power classes 3 and 4 the average transmit output power must be 20 dBm and 18 dBm, respectively. ACS is defined as the ratio of the receive filter attenuation on the wanted channel frequency to the receive filter attenuation on the adjacent channel and must be better than the defined 33 dB for power classes 3 and 4. As the specified power of $-103\,\text{dBm}/3.84\,\text{MHz}$ for the DPCH is 14 dB above the sensitivity limit, noise is of no concern in this test case. The I_{oac} (modulated) signal is specified [105] to consist of the common channels needed for the receiver tests [116] and 16 dedicated data channels. The channelization codes for data channels should be chosen optimally to reduce the peak-to-average ratio, and the user data should be uncorrelated. If we again assume a combined spreading and coding gain of 25 dB, the ACS requirements suppress the adjacent channel power at least down to the required $-7.2\,\text{dB}$ relative to the despread and decoded signal energy as is shown in Fig. 6.38.

In-band Blocking. The in-band blocking test case is similar to the adjacent channel selectivity test case, but here the unwanted signals with much higher power than the wanted receive signal are 10 and 15 MHz away. The parameters for this test case are given in Table 6.34. For power classes 3 and 4, the average output power of the simultaneously operating transmitter must be 20 dBm and 18 dBm, respectively. As in the ACS test case, the blocking signal $I_{blocking}$ (modulated) consists of the necessary common channels for any connection and 16 dedicated data channels with uncorrelated user data. Again, the channelization codes for data channels are chosen optimally to reduce the peak-to-average ratio. A graphical representation of the test case is shown in Fig. 6.39.

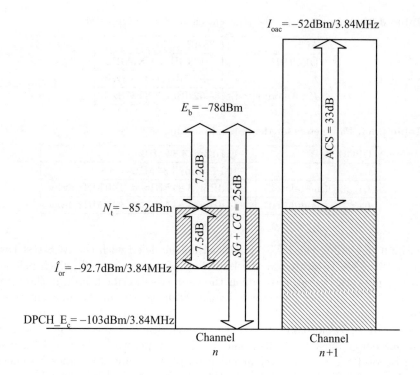

Fig. 6.38. Adjacent channel selectivity test case

From the in-band blocking test case, a requirement for the IP2 for direct-conversion receivers can be derived. If we repeat the calculation from the reference sensitivity test case, we find the acceptable interference level N_t to be

$$
\begin{aligned}
N_t &= P_{\text{DPCH}} + G_p + G_c - \left(\frac{E_b}{N_t}\right)_{\text{eff}} \\
&= -114\,\text{dBm} + 21\,\text{dB} + 4\,\text{dB} - 7.2\,\text{dB} \\
&= -96.2\,\text{dBm}.
\end{aligned}
\tag{6.19}
$$

Here N_t is the total noise plus interference power with the interference resulting *only* from the second-order nonlinearity. Because the power of the desired signal is only 3 dB above the reference sensitivity level, we assume an equal distribution of N_t between noise (P_N) and interference ($P_{\text{I},2}$) resulting in $P_{\text{I},2} = -99.2\,\text{dBm}$. Owing to the squaring of the modulated blocker signal, half of the interference power appears at DC and the other half of the power appears in a 10 MHz bandwidth [119]. If we assume that high- and low-pass filtering is performed in the receiver, which removes the DC component and the interference above the channel bandwidth, an improve-

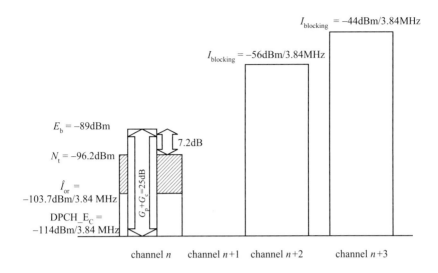

Fig. 6.39. In-band blocking test case

ment of about 6 dB results. Using (2.60), assuming $G = 0$, and setting $P_{1,2} = -99.2 + 6 = -93.2$ dBm, we find the required IIP2 to be

$$\text{IIP2} = 2P_{\text{in}} - P_{1,2} = 2 \cdot (-44) \text{ dBm} + 93.2 \text{ dBm} = 5.2 \text{ dBm}. \qquad (6.20)$$

A much more stringent IIP2 requirement results if we consider the ever-present transmitter leakage signal at the receiver input. We assume a -30 dBm transmitter leakage signal at the receiver input and require the interference level from the second-order nonlinearity to be 10 dB below the noise level (since the transmitter leakage signal is always present).

$$\text{IIP2} = 2P_{\text{in}} - P_{1,2} = \qquad\qquad (6.21)$$
$$2 \cdot (-30) \text{ dBm} - (-93.2 - 10) \text{ dBm} = 43.2 \text{ dBm}.$$

Such a calculation of the IIP2 requirement is only a coarse estimation. What has been neglected in this calculation is the fact that the IIP2 calculation according to (2.60) is based on pure sinusoidal signals and all signals occurring in the in-band blocking test case are W-CDMA signals. Furthermore, the assumptions of 6 dB improvement, because of high- and low-pass filtering and of the required interference levels, are only estimations.

Out-of-Band Blocking. In the out-of-band blocking test case, the BER must not exceed 10^{-3} in the presence of CW (continuous wave) interferers outside the receive band. The specifications are listed in Table 6.35 and a graphical representation is given in Fig. 6.40 for operation in countries outside ITU region 2. For power classes 3 and 4 the average output power of the simultaneously operating transmitter has to be 20 dBm and 18 dBm, respectively. For frequencies in the ranges 2095 MHz $< f <$ 2110 MHz and

Table 6.35. Parameters for the out-of-band blocking test case

DPCH$\_E_C$		$-114\,\mathrm{dBm}/3.84\,\mathrm{MHz}$
$\hat{I}_{\mathrm{or}}$		$-103.7\,\mathrm{dBm}/3.84\,\mathrm{MHz}$
I_{blocking} (CW)	$-44\,\mathrm{dBm}$	$2050\,\mathrm{MHz} < f < 2095\,\mathrm{MHz}$ $2185\,\mathrm{MHz} < f < 2230\,\mathrm{MHz}$
$I_{\mathrm{blocking}}{}^{1}$ (CW)	$-44\,\mathrm{dBm}$	$1870\,\mathrm{MHz} < f < 1915\,\mathrm{MHz}^{1}$ $2005\,\mathrm{MHz} < f < 2050\,\mathrm{MHz}^{1}$
I_{blocking} (CW)	$-30\,\mathrm{dBm}$	$2025\,\mathrm{MHz} < f < 2050\,\mathrm{MHz}$ $2230\,\mathrm{MHz} < f < 2255\,\mathrm{MHz}$
$I_{\mathrm{blocking}}{}^{1}$ (CW)	$-30\,\mathrm{dBm}$	$1845\,\mathrm{MHz} < f < 1870\,\mathrm{MHz}^{1}$ $2050\,\mathrm{MHz} < f < 2075\,\mathrm{MHz}^{1}$
I_{blocking} (CW)	$-15\,\mathrm{dBm}$	$1\,\mathrm{MHz} < f < 2025\,\mathrm{MHz}$ $2255\,\mathrm{MHz} < f < 12750\,\mathrm{MHz}$
$I_{\mathrm{blocking}}{}^{1}$ (CW)	$-15\,\mathrm{dBm}$	$1\,\mathrm{MHz} < f < 1845\,\mathrm{MHz}^{1}$ $2075\,\mathrm{MHz} < f < 12750\,\mathrm{MHz}^{1}$

[1] If operated in ITU Region 2, which comprises the countries of North and South America.

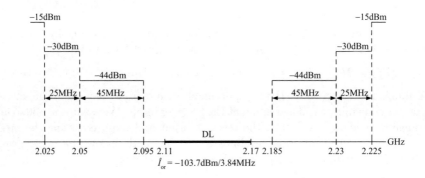

$\hat{I}_{\mathrm{or}} = -103.7\,\mathrm{dBm}/3.84\,\mathrm{MHz}$

Fig. 6.40. Out-of-band blocking test case for countries outside region 2

$2170\,\mathrm{MHz} < f < 2185\,\mathrm{MHz}$, the appropriate in-band blocking or adjacent channel selectivity test case specification must be fulfilled. For operation in ITU region 2, this is valid for the frequency ranges $1915\,\mathrm{MHz} < f < 1930\,\mathrm{MHz}$ and $1990\,\mathrm{MHz} < f < 2005\,\mathrm{MHz}$. For this test case, up to 24 exception frequencies are allowed in each assigned frequency channel (when measured using a 1 MHz step size) at which spurious response signals occur and the out-of-band blocking specifications need not be fulfilled. At these exception frequencies the spurious response test case must be met (see later).

From the out-of-band blocking test case a requirement for the receiver's IIP3 can be derived. At the receiver input we assume a transmitter leakage signal of, e.g., $-30\,\mathrm{dBm}$ (due to limited duplexer attenuation [119]) at $1930\,\mathrm{MHz}$ and a blocker signal with $-15\,\mathrm{dBm}$ at $2025\,\mathrm{MHz}$, which is half the

Table 6.36. Parameters for the intermodulation test case

DPCH_E_C	$-114\,\text{dBm}/3.84\,\text{MHz}$
$\hat{I}_{\text{or}}$	$-103.7\,\text{dBm}/3.84\,\text{MHz}$
I_{ouw1} (CW)	$-46\,\text{dBm}$ at $\pm10\,\text{MHz}$ away
I_{ouw2} (modulated)	$-46\,\text{dBm}/3.84\,\text{MHz}$ at $\pm20\,\text{MHz}$ away

duplex distance if we assume a fixed duplex distance of 190 MHz. In this case any third-order nonlinearity produces a spurious signal at 2120 MHz, which occurs at the same frequency as the received signal owing to the fixed duplex distance of 190 MHz. The acceptable noise plus interference level N_t (here the interference results *only* from the third-order nonlinearity) is found to be $-96.2\,\text{dBm}$, following the same calculation as in the in-band blocking test case (6.19). If we again assume an equal distribution of N_t between noise and interference, we arrive at $-99.2\,\text{dBm}$ interference power P_3. Using (2.59) with $P_1 = -30\,\text{dBm}$ (assuming a blocker suppression of 15 dB by the duplexer), $P_2 = -30\,\text{dBm}$ (the transmitter leakage signal), and $G = 0$, we find for the required IIP3

$$\text{IIP3} = \frac{2P_1 + P_2 - P_3}{2} = 4.6\,\text{dBm}. \tag{6.22}$$

It must be noted, however, that this type of calculation of the IIP3 requirement can only serve as a rough estimation. What has been neglected in this calculation is the fact that the IIP3 calculation according to (2.59) is based on pure sinusoidal signals. The IIP3 of a nonlinear building block with respect to a W-CDMA signal is different from the IIP3 for sinusoidal signals [47]. A further point neglected in the above estimation is the fact that the transmit signal is a W-CDMA signal spread with the same type of OVSF codes as the received signal (but the actual spreading codes used in the transmit and the received signal will in general be different, and the scrambling codes are also different).

Intermodulation Test Case. For the intermodulation test case two types of interferers are specified: a CW interferer and a modulated interference signal. The parameters are given in Table 6.36. For power classes 3 and 4, the average output power of the simultaneously operating transmitter must be 20 dBm and 18 dBm, respectively. As for the in-band blocking test case, the modulated interference signal I_{ouw2} (modulated) consists of the necessary common channels for any connection and 16 dedicated data channels with uncorrelated user data. The channelization codes for the data channels are chosen optimally to reduce the peak-to-average ratio. Figure 6.41 depicts the test case conditions. The sum of both signals is transferred on a third-order nonlinearity into the desired channel. Therefore, this test case defines a requirement for the IIP3 of the receiver. As for the in-band blocking case, the acceptable noise plus interference level N_t in the desired channel must not

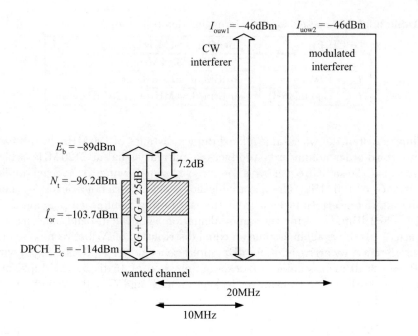

Fig. 6.41. Intermodulation test case

exceed $-96.2\,\mathrm{dBm}/3.84\,\mathrm{MHz}$, if we assume a combined spreading and coding gain of 25 dB. Before determining the required IIP3 we must distribute the total noise and interference power N_t to their sources. According to [119], we assume the following: 50% noise power $(-3\,\mathrm{dB})$, 15% intermodulation power $(-8\,\mathrm{dB})$, 15% blocking from the CW interferer$(-8\,\mathrm{dB})$, 15% blocking from the modulated interferer$(-8\,\mathrm{dB})$, and 5% power from oscillator noise $(-13\,\mathrm{dB})$. We neglect second order products. From our assumptions we derive a tolerable level for third order intermodulation power of $P_3 = -96.2 - 8 = -104.2\,\mathrm{dBm}/3.84\,\mathrm{MHz}$. If we repeat the IIP3 computation from the out-of-band blocking test case we find

$$\mathrm{IIP3} = \frac{2I_{\mathrm{ouw},1} + I_{\mathrm{ouw},2} - P_3}{2} = -16.9\,\mathrm{dBm} \qquad (6.23)$$

with $I_{\mathrm{ouw},1} = I_{\mathrm{ouw},2} = -46\,\mathrm{dBm}$. This requirement for the IIP3 is far more relaxed than that derived from the out-of-band blocking test case. Again, this computation only serves as a rough estimation.

Spurious Response Test Case. At the exception frequencies of the out-of-band blocking test case (where spurious response signals occur) the specifications listed in Table 6.37 must be met.

Spurious Emissions. The spurious emissions generated by the MS receiver at the antenna connector must be less than

Table 6.37. Parameters for the spurious response test case

DPCH_E_C	$-114\,\text{dBm}/3.84\,\text{MHz}$
$\hat{I}_{\text{or}}$	$-103.7\,\text{dBm}/3.84\,\text{MHz}$
I_{blocking} (CW)	$-44\,\text{dBm}$ at spurious response frequencies

Information bit rate	12.2 kbps
DPCH data rate	60 kbps
Spreading factor DPDCH	64
Spreading factor DPCCH	256
Power control	Off
TFCI	On
Pilot bits/slot	6
Power control bits/slot	2
TFCI bits/slot	2
Power ratio of DPCCH/DPDCH	$-2.69\,\text{dB}$

Table 6.38. Physical parameters for the 12.2 kbps UL reference measurement channel

- $-60\,\text{dBm}/3.84\,\text{MHz}$ in the DL band and for certain operation modi (URA_PCH-, Cell_PCH-, and IDLE- stages) [105] also in the UL band,
- $-57\,\text{dBm}/100\,\text{kHz}$ for $9\,\text{kHz} < f < 1\,\text{GHz}$, and
- $-47\,\text{dBm}/1\,\text{MHz}$ for $1\,\text{GHz} < f < 12.75\,\text{GHz}$.

6.8.5 BS Receiver Test Cases

For the BS receiver test cases it is assumed that the receiver does not apply diversity. For receivers with diversity, the requirements apply to each antenna connector separately, with the other one(s) terminated or disabled. Unless otherwise stated all parameters are specified at the antenna connector of the BS. They are defined using the 12.2 kbps UL reference measurement channel with the specifications listed in Table 6.38. The coded bit error rate under the conditions stated for each receiver test case must not exceed 10^{-3}.

Reference Sensitivity Test Case. The BS reference sensitivity level is specified to be $-121\,\text{dBm}$.

Dynamic Range. The dynamic range of the receiver is its ability to handle a rise of interference in the reception frequency channel. The interference is assumed to be AWGN. The BER of 10^{-3} must be fulfilled for a wanted signal level of $-91\,\text{dBm}$ in the presence of an interferer level of $-73\,\text{dBm}/3.84\,\text{MHz}$.

Adjacent Channel Selectivity Test Case. The parameters for this test case are listed in Table 6.39.

Table 6.39. Parameters for the BSs adjacent channel selectivity test case

Wanted signal	$-115\,\mathrm{dBm}$
Interfering signal	$-52\,\mathrm{dBm}/3.84\,\mathrm{MHz}$
Frequency offset	5 MHz

Blocking Characteristics. The blocking characteristics describes the ability of the BS to receive the wanted signal in the presence of an interferer on frequencies other than those of the adjacent channels. The blocking characteristics must be fulfilled at all frequencies listed in Table 6.40, using a 1 MHz step size.

Table 6.40. Parameters for the blocking characteristic

Center frequency of interfering signal	Interfering signal level	Wanted signal level	Minimum offset of interferer	Type of interfering signal
$1920 - 1980\,\mathrm{MHz}$	$-40\,\mathrm{dBm}$	$-115\,\mathrm{dBm}$	10 MHz	W-CDMA signal with one code
$1850 - 1910\,\mathrm{MHz}$ [1]	$-40\,\mathrm{dBm}$	$-115\,\mathrm{dBm}$	10 MHz	W-CDMA signal with one code
$1900 - 1920\,\mathrm{MHz}$ $1980 - 2000\,\mathrm{MHz}$	$-40\,\mathrm{dBm}$	$-115\,\mathrm{dBm}$	10 MHz	W-CDMA signal with one code
$1830 - 1850\,\mathrm{MHz}$ $1910 - 1930\,\mathrm{MHz}$ [1]	$-40\,\mathrm{dBm}$	$-115\,\mathrm{dBm}$	10 MHz	W-CDMA signal with one code
$1 - 1900\,\mathrm{MHz}$ $2000 - 12750\,\mathrm{MHz}$	$-15\,\mathrm{dBm}$	$-115\,\mathrm{dBm}$	–	CW
$1 - 1830\,\mathrm{MHz}$ $1930 - 12750\,\mathrm{MHz}$ [1]	$-15\,\mathrm{dBm}$	$-115\,\mathrm{dBm}$	–	CW
$925 - 960\,\mathrm{MHz}$ [2]	$+16\,\mathrm{dBm}$	$-115\,\mathrm{dBm}$	–	CW
$1805 - 1880\,\mathrm{MHz}$ [3]	$+16\,\mathrm{dBm}$	$-115\,\mathrm{dBm}$	–	CW

[1] If operated in ITU region 2, which comprises the countries of North and South America.
[2] Only applicable outside region 2 if the BS is co-located with a GSM900 base transceiver station.
[3] Only applicable outside region 2 if the BS is co-located with a DCS1800 base transceiver station.

Intermodulation Test Case. For the intermodulation test case, two types of interferers are specified: a CW interferer and a modulated interference

Table 6.41. Parameters for the BSs intermodulation test case

Wanted signal	$-115\,$dBm
CW interferer	$-48\,$dBm at $10\,$MHz away
W-CDMA signal with one code	$-48\,$dBm at $20\,$MHz away

signal. The parameters are given in Table 6.41. Both interfering signals are fed into the BS antenna input together with the wanted signal. Under these conditions the BER must stay below 10^{-3}.

Spurious Emissions. The spurious emissions generated by the BS receiver at the antenna connector must be less than

- $-78\,$dBm$/3.84\,$MHz for $1900 - 1980\,$MHz and $2010 - 2025\,$GHz,
- $-57\,$dBm$/100\,$kHz for $9\,$kHz $< f < 1\,$GHz, and
- $-47\,$dBm$/1\,$MHz for $1\,$GHz $< f < 12.75\,$GHz.

The requirements apply to all BSs with separate receive and transmit antenna ports and must be tested if both transmitter and receiver are operating, and with the transmit port terminated.

7. UTRA TDD Mode

The UTRA TDD mode is described in this chapter. We will concentrate on the differences to the FDD mode and will not repeat specifications which are the same as for the FDD mode.

7.1 Key Parameters

Table 7.1 lists the key parameters of the UTRA TDD mode. A detailed description of these parameters will be given in the remainder of this chapter. In UTRA TDD mode, two options for the chip rate are possible, namely 3.84 Mcps and 1.28 Mcps. The 1.28 Mcps option was proposed by the Chinese Wireless Telecommunications Standards organization to realize integration with the current TDD standards [122]. The associated channel spacing of 1.6 MHz fits three 1.28 Mcps channels into one 3.84 Mcps channel.

7.2 Transport and Physical Channels for the 3.84 Mcps Option

While the physical format is different, most of the transport and physical channels are the same in FDD and TDD mode. Table 7.2 lists the transport and physical channels and the mapping between them for the 3.84 Mcps option [123]. If listed channels are not described in the following there are no relevant differences to the FDD mode in the physical layer.

In the 3.84 Mcps option of the UTRA TDD mode, the frame and slot structure is the same as in the FDD mode and is shown in Fig. 7.1. The main distinctions are that the timeslots can be used either for UL *or* DL and that users can be separated in the time domain by assigning different users to different timeslots. This separation is performed *in addition* to the separation in the code domain, which is the same is in FDD.

Each of the 15 slots of length 0.666 ms in a frame can be either allocated to the DL or the UL. The only applicable restriction is that at least one timeslot must be assigned each for DL and UL. The switching point(s) between UL and DL can be chosen arbitrarily. Examples are shown in Fig. 7.2. With this

Table 7.1. Key parameters of the UMTS TDD mode

Parameter	Value	Description
Band A	1900–1920 MHz	UL and DL transmission
Band A	2010–2025 MHz	UL and DL transmission
Band B	1850–1910 MHz	UL and DL transmission in region 2[1]
Band B	1930–1990 MHz	UL and DL transmission in region 2[1]
Band C	1910–1930 MHz	UL and DL transmission in region 2[1]
f_{chip}	3.84 Mcps 1.28 Mcps	Chip rate (2 options)
Δf	5 MHz 1.6 MHz	Channel spacing (3.84 Mcps option) (1.28 Mcps option)
T_{chip}	0.26042 μs 0.78125 μs	Duration of a chip (3.84 Mcps) (1.28 Mcps)
N_{chip}	2560 864	Number of chips per slot (3.84 Mcps) (1.28 Mcps)
T_{slot}	0.666 ms 0.675 ms	Duration of a slot (3.84 Mcps) (1.28 Mcps)
T_{frame}	10 ms	Duration of a frame
$T_{superframe}$	720 ms	Duration of a superframe
$f_{bit,inf}$	12.2..2048 kbit/s	Information bit rate
$f_{bit,ph}$	24.4..441.6 kbps 8.8..211.2 kbps	Physical bit rate with 3.84 Mchip/s and with 1.28 Mchip/s
BS synchronization	required	
SF(UL/DL)	16..1	Spreading factor
α	0.22	Roll-off factor (root-raised cosine)

---

[1] ITU Region 2 comprises the countries of North and South America.

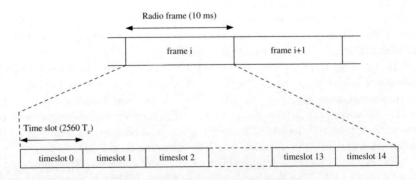

Fig. 7.1. Frame and slot structure of the physical channels in the TDD mode for the 3.84 Mcps option

Table 7.2. Transport and physical channels and their mapping for the 3.84 Mcps option of the UTRA TDD mode

Transport Channel		Physical Channel	
Dedicated Uplink Channels			
Dedicated Channel	DCH	Dedicated Physical Channel	DPCH
Dedicated Downlink Channels			
Dedicated Channel	DCH	Dedicated Physical Channel	DPCH
Common Uplink Channels			
Random Access Channel	RACH	Physical Random Access Channel	PRACH
Uplink Shared Channel	USCH	Physical Uplink Shared Channel	PUSCH
Common Downlink Channels			
Broadcast Channel	BCH	Primary Common Control Physical Channel	P-CCPCH
Forward Access Channel	FACH	Secondary Common Control Physical Channel	S-CCPCH
Paging Channel	PCH	Secondary Common Control Physical Channel	S-CCPCH
Downlink Shared Channel	DSCH	Physical Downlink Shared Channel	PDSCH
		Synchronization Channel	SCH
		Paging Indicator Channel	PICH

high flexibility the adaptation to various kinds of environments and types of data traffic (especially to highly asymmetric traffic) becomes possible. The physical channels in TDD are further mapped to bursts, which are the signals transmitted in a particular timeslot. Therefore, a physical channel is defined by its frequency, timeslot, channelization code, burst type and radio frame allocation. The radio frame allocation is continuous if the channel is transmitted in the same timeslot in every frame. In the case of discontinuous allocation the channel is transmitted only in certain frames. It is possible to transmit several bursts at the same time from one BS or MS. In this case different OVSF codes must be used for the different bursts.

7.2.1 Dedicated Channel

The DCH is mapped onto the DPCH, which is transmitted using bursts. Each burst consists of a data part, a midamble, and a guard period. The data part of the burst is spread with the same OVSF codes as used in the UTRA FDD mode (see section 6.4.1). In TDD, only SFs of 1, 2, 4, 8, and 16 are used. Scrambling is performed after spreading (for further details see section 7.4.3). Three types of bursts for the DPCHs are defined, depending

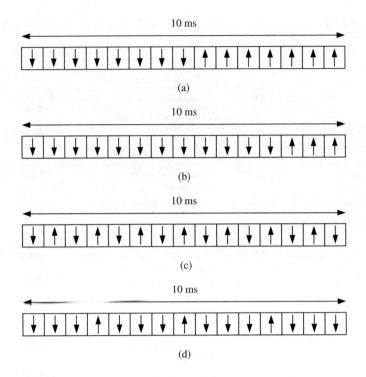

Fig. 7.2. Switching point configurations. (a) Single switching point configuration with symmetric DL/UL allocation (b) Single switching point configuration with asymmetric DL/UL allocation (c) Multiple switching point configuration with symmetric DL/UL allocation (d) Multiple switching point configuration with asymmetric DL/UL allocation

on the length of the midamble and the guard period. *Burst type 1* has a midamble with a length of 512 chips. The structure of this burst is shown in Fig. 7.3a. The midamble is located in between the two data fields with length 976 chips each. The guard period has a length of 96 chips corresponding to a guard time of 25 µs. In *burst type 2* the midamble has a length of 256 chips resulting in two 1104 chip long data fields, since the guard period is the same as with burst type 1 (Fig. 7.3b). *Burst type 3* has the same midamble length of 512 chips as burst type 1 but a guard period of 192 chips, leaving 880 chips for the second data part while the first data part is unchanged with 976 chips. Figure 7.3c shows the structure of this burst. Table 7.3 summarizes the number of symbols per data field for the three burst types. In the same timeslot of one cell, burst type 2 must not be mixed with burst types 1 or 3.

The midamble, which is known to the receiver, is used for channel estimation and coherent detection. Because of the longer midamble, burst type 1 is

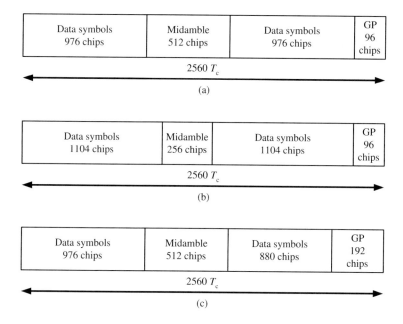

Fig. 7.3. Structure of (a) burst type 1, (b) burst type 2, and (c) burst type 3 used for the 3.84 Mcps TDD option

Table 7.3. Number of data symbols in burst type 1, 2, and 3

SF	Number of data symbols		
	Burst type 1	Burst type 2	Burst type 3
1	1952	2208	1856
2	976	1104	928
4	488	552	464
8	244	276	232
16	122	138	116

well suited for the UL, where up to 16 different channel impulse responses can be estimated in the BS. Burst type 1 can also be used in the DL. Burst type 2 is used for the DL, too. It can be used in the UL if the bursts within a timeslot are allocated to less than four users [75]. Burst type 3 is used in UL only, since with the longer guard period it is well suited for initial access or access to a new cell after handover. Each user in a timeslot has its own midamble which is a cyclically shifted version of one single basic midamble code. For each burst type, 127 basic midamble codes exist. They are defined in [123] together with the rules concerning extraction of the specific midambles. The midambles are not spread or scrambled.

Transport format combination indicator bits, which indicate the combination of used transport channels on the DPCH, can be transmitted with all three burst types. The transmission of the TFCI bits is negotiated at call setup and can be re-negotiated during the call. The TFCI appears always in the first timeslot in a radio frame for each CCTrCH. The TFCI is transmitted in the data parts of the respective physical channel. Therefore, it is subject to spreading, and the midamble structure and length are not changed. In the UL the TFCI information is always spread with a SF of 16 using the channelization code in the lowest branch of the allowed OVSF sub tree. In the DL, a SF of 16 is fixed for the data part (see section 7.4.2), which is also used for the TFCI. The TFCI information is split into two parts with the first part being transmitted directly before and the second part directly after the midamble.

A feature of all three burst types is that the provision of *transmit power control* information in the UL is possible. The transmission of TPC is negotiated at call setup and can be re-negotiated during the call. As with the TFCI bits, the data part is used to carry the TPC information and thus the midamble is not changed. The TPC information is transmitted directly after the midamble. If both TPC and TFCI information is transmitted in the same slot, the TPC bits are inserted between the midamble and the second part of the TFCI, i.e., they are transmitted in the same timeslot. Furthermore, for the TPC the same channelization code is used as for the TFCI information. Otherwise the TPC bits are transmitted in the first allocated timeslot and with the first allocated channelization code, according to the order in the higher layer allocation message. For every user the TPC bits are transmitted at least once per radio frame. The spreading of the TPC is always performed with SF = 16, independent of the spreading of the data part. The burst structure for both burst types with and without TPC transmission is shown in Fig. 7.4.

7.2.2 Random Access Channel

The RACH, a common UL transport channel used to carry control information and short user data packets, is mapped onto the PRACH. The PRACH is transmitted with burst type 3 because of its longer guard period of 192 chips (50 µs) which corresponds to a cell radius of 7.5 km. This is necessary as the required timing advance value in systems with a TDMA component is not known at the time the MS requests a connection setup via the RACH. No TFCI and TPC information is transmitted on the PRACH. Two types of slot formats are possible. One uses a SF of 16 resulting in 232 bits per slot and the other format employs a SF of 8 with 464 bits in one slot. For the PRACH a fixed association between the training sequence and the channelization code exists, which is described in detail in [123].

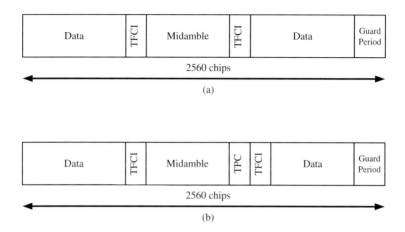

Fig. 7.4. General burst structure for traffic burst types 1, 2, and 3 (a) without TPC and (b) with TPC transmission

7.2.3 Broadcast Channel

The BCH, a common DL transport channel used to broadcast system- and cell-specific information, is mapped onto the P-CCPCH. For the P-CCPCH burst type 1 with a fixed SF of 16 and without TFCI is used.

7.2.4 Paging Channel and Forward Access Channel

The PCH is a common DL transport channel used to transmit control information to a MS if the system does not know the location cell of the MS. If the UTRAN knows the location cell of the MS, the FACH is used instead. The FACH can also be used to transmit short user data packets. Both FACH and PCH are mapped onto one or more S-CCPCHs to adapt the capacity of PCH and FACH to different requirements. The S-CCPCHs are transmitted using burst types 1 or 2 with a fixed SF of 16. TFCI information can be transmitted on the S-CCPCH.

7.2.5 Synchronization Channel

As in the FDD mode, the SCH provides synchronization and the code group information to the MS. The SCH is mapped only on one or two DL slots per frame to allow for maximum UL/DL asymmetry for the data traffic. For the SCH allocation two cases can be distinguished:

- *Case 1*: SCH is allocated to slot number $k \in \{0 \ldots 14\}$

- *Case 2*: SCH is allocated to slot number k and $k + 8$ with $k \in \{0 \dots 6\}$.

In both cases the P-CCPCH is transmitted in parallel in slot number k. Thus, its position is known if the SCH has been detected. The SCH consists of a primary and three secondary code sequences each with a length of 256 chips (see section 7.4.5).

In public TDD systems it is mandatory to synchronize the base stations to limit the interference between MSs. Therefore, it is possible that a MS can detect the SCH of two neighboring cells in the same slot. This is called capturing effect. In UTRA TDD, a time offset t_{offset} between the slot timing and the SCH is introduced to overcome this capture effect. The timing offset between different cells is given by

$$
t_{\text{offset},n} = \begin{cases} 48 n T_{\text{chip}} & n < 16 \\ (720 + n48)\, T_{\text{chip}} & n \geq 16 \end{cases} \; ; \quad n = 0, \dots, 31 . \tag{7.1}
$$

The actual value of n depends on the cell parameters and is encoded in the SCH. Figure 7.5 schematically shows the timing of the SCH consisting of one primary and three secondary code sequences for case 2 with $k = 0$.

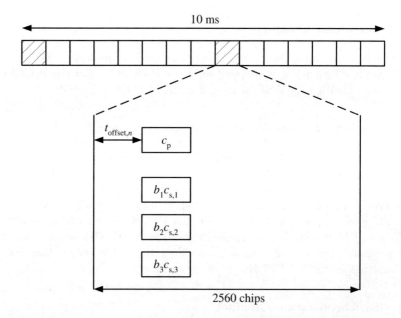

Fig. 7.5. Scheme for the SCH consisting of one primary and three secondary codes for case 2 with $k = 0$

7.2.6 Uplink and Downlink Shared Channel

The UL and DL Shared Channels (USCH/DSCH), which are mapped to one or several of the corresponding UL and DL Physical Shared Channels (PUSCH/PDSCH), are transport channels shared by several MSs carrying dedicated control or traffic data. All three burst types can be used with the PUSCH, while the PDSCH uses only burst types 1 and 2.

7.2.7 Paging Indicator Channel

The PICH is a physical channel used to carry the Paging Indicators (PI). The PI indicates a paging message for one or more MSs that are associated with it. The PICH is transmitted with burst type 1 or 2. Details on the bit structure of the data parts carrying the PI information can be found in [123].

7.2.8 The Physical Node B Synchronization Channel

So-called "cell sync bursts" can be used for BS (Node B) synchronization. The sync bursts are transmitted on the physical node B synchronization channel (PNBSCH), which is mapped on the same timeslot as the PRACH. The cell sync burst is transmitted at the beginning of a timeslot. It contains of 2304 chips and a guard period of 256 chips.

7.3 Transport and Physical Channels for the 1.28 Mcps Option

In this section only the differences between the 3.84 Mcps and the 1.28 Mcps options are described. The transport and physical channels in the TDD 1.28 Mcps option are the same as in the 3.84 Mcps option with the following exceptions:

- The SCH and the PICH do not exist.
- The following three physical channels appear, which have no transport channel mapped to them: Downlink Pilot Channel (DwPCH), Uplink Pilot Channel (UpPCH), and Fast Physical Access Channel. (FPACH)

Figure 7.6 shows the frame and slot structure of the physical channels in the 1.28 Mcps option of the TDD mode. As opposed to the 3.84 Mcps option, each radio frame is subdivided into two sub-frames, each with a duration of 5 ms. Furthermore, the structure in each sub-frame, shown in Fig. 7.7, differs from the 3.84 Mcps option. The chip duration is $T_{\text{chip}} = 781.25$ ns, and each timeslot consists of 864 chips yielding $T_{\text{slot}} = 675$ µs. Timeslot 0 and 1 are always allocated for DL and UL, respectively. Inserted between timeslot 0 and 1 are the Downlink Pilot Timeslot (DwPTS, 96 chips), the main guard

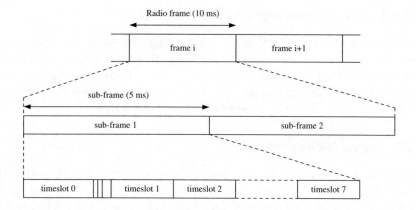

Fig. 7.6. Frame and slot structure of the physical channels in the TDD mode for the 1.28 Mcps option

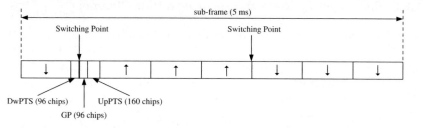

Fig. 7.7. Structure of the sub-frame for the 1.28 Mcps TDD option

period (GP, 96 chips), and the Uplink Pilot Timeslot (UpPTS, 160 chips). Details on the channels to be transmitted in the DwPTS and UpPTS are described in section 7.5.2. In each sub-frame there are two switching points (from DL to UL and vice versa). The second switching point in the sub-frame can be chosen anywhere after timeslot 1.

7.3.1 Dedicated Channel

The structure of the burst used to transmit the DPCH in the 1.28 Mcps TDD option is shown in Fig. 7.8. It consists of two data parts each 352 chips long, separated by a 144 chip long midamble and ends with a guard period of 16 chips. The total number of data symbols in one slot ranges from 44 with a SF of 16 to 704 with a SF of 1.

The transmission of TFCI, TPC, and Synchronization Shift (SS) commands is possible in UL and DL. The TFCI symbols are distributed between the data parts of the two sub-frames. For every user the TPC information must be transmitted at least once per sub-frame. If applied, the TPC is transmitted in the data parts of the burst and it can be transmitted using the first

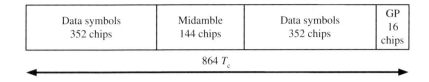

Fig. 7.8. Structure of the burst used for the 1.28 Mcps TDD option

allocated channelization code and the first allocated timeslot. Other allocations (more than one TPC transmission in one sub-frame) of the TPC bits are also possible. TFCI and TPC bits are spread with the same SF and spreading code as the data parts of the respective physical channel. The transmission of an uplink synchronization control (ULSC) information is possible by means of SS commands. The ULSC information must be transmitted at least once per sub-frame for every user. The SS bits are transmitted in the data parts of the burst. Hence, the midamble structure and length are not changed. The SS information is transmitted directly after the midamble. Figure 7.9 shows

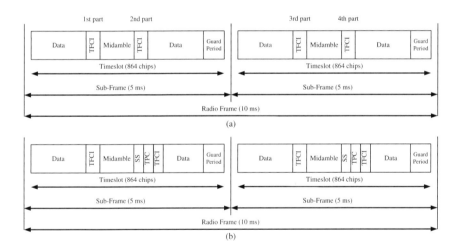

Fig. 7.9. Position of TFCI if (a) no SS and TPC and (b) if SS and TPC are transmitted

the positions of the TFCI, TPC and SS information in a traffic burst. In the UL no SS information is transmitted. However, the SS position in the bursts is reserved for future use, which gives an identical slot structure in UL and DL. Details on the TPC and SS information is provided in [123].

7.3.2 Broadcast Channel

The BCH, a common DL transport channel used to broadcast system- and cell-specific information, is mapped onto the P-CCPCHs (P-CCPCH1 and P-CCPCH2). The P-CCPCHs are mapped onto the first two code channels of timeslot 0 with a SF of 16. The P-CCPCH is always transmitted with an antenna pattern configuration that provides whole cell coverage. The P-CCPCH1 and P-CCPCH2 always use the channelization codes $c_{Q=16}^{(k=1)}$ and $c_{Q=16}^{(k=2)}$, respectively. The midamble $m(1)$ is used for timeslot 0.

7.3.3 Random Access Channel

The UL PRACH which carries the RACH can use a SF of 16, 8, or 4.

7.3.4 Fast Physical Access Channel

The FPACH, transmitted by the BS, is used during the random access procedure (see section 7.7.8). It acknowledges the detection of a signature (SYNCH-UL code, see sections 7.3.5 and 7.5.2) in a single burst with timing and power level adjustment indication. The FPACH uses only a SF of 16.

7.3.5 The Synchronization Channels (DwPCH, UpPCH)

In the 1.28 Mcps TDD option, two dedicated physical synchronization channels appear: the DwPCH and the UpPCH. DwPCH and UpPCH are transmitted during the DwPTS and UpPTS, respectively. They both appear in each sub-frame (see Fig. 7.7). The burst structures of the DwPCH (DwPTS) and the UpPCH (UpPTS) are shown in Fig. 7.10. The DwPCH contains a 32-chip long guard period and the SYNCH-DL sequence with a length of 64 chips. On the UpPCH the SYNCH-UL sequence (= signature) consisting of 128 chips and a guard period with 32 chips are transmitted (see section 7.5.2).

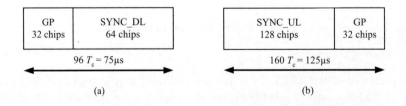

Fig. 7.10. Burst structure of the (a) DwPCH (DwPTS) and the (b) UpPCH (Up-PTS)

7.4 Spreading and Modulation for the 3.84 Mcps Option

As in the FDD mode, spreading and scrambling is applied to the data carried on the physical channels in TDD mode. While the purpose of these operations is the same in both modes, the details are slightly different and are described in detail in [124].

7.4.1 QPSK Mapping of the Data Bits

Different to the FDD mode, the interleaved and encoded (see section 7.6) data bits are mapped to a QPSK constellation before spreading. The mapping is performed according to Table 7.4. Here, the superscript $i = 1, 2$ distinguishes

Table 7.4. Mapping of the data bits b to QPSK symbols $\underline{d}$

Consecutive binary bit pattern $b_{1,n}^{(k,i)} b_{2,n}^{(k,i)}$	Complex symbol $\underline{d}_n^{(k,i)}$
00	$+j$
01	$+1$
10	-1
11	$-j$

between the two data fields of the traffic bursts, $k = 1, \ldots, K$ denotes the kth user, and the subscript $n = 1, \ldots, N_k$ identifies the nth data symbol out of a total of N_k symbols per data field for the user k. This QPSK mapping is the same for all three types of traffic bursts.

7.4.2 Channelization

The spreading of the complex data symbols to the chip rate of 3.84 Mcps is performed using the OVSF codes introduced in section 6.4.1, Fig. 6.6. The complex data symbols $\underline{d}_n^{(k,i)}$ are multiplied with the real valued channelization codes $\mathbf{c}^{(k)} = (c_1^{(k)}, c_2^{(k)}, \ldots, c_{Q_k}^{(k)})$; $k = 1, \ldots, K$, which take values out of $\{+1, -1\}$. Q_k denotes the SF. Because only mutual orthogonal OVSF codes are allowed in one timeslot, all bursts sent simultaneously in one timeslot can be separated by the despreading operation in the receiver. SFs of 1 to 16 are specified for the UL physical channels. In the case of multi-code transmission not more than two physical channels per timeslot can be used simultaneously. In the DL, the SF is fixed at 16 while multiple physical channels can be transmitted in parallel. If only a single channel is transmitted in a timeslot, SF $= 1$ (no spreading) is allowed.

A multiplier $w_{Q_k}^{(k)}$ is associated with each channelization code and can take values out of the set $\{\exp(j\pi/2p_k)\}$ with p_k being a permutation of the

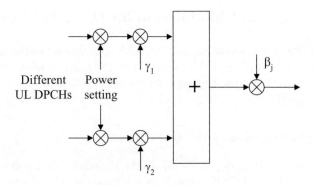

Fig. 7.11. Combination of different DPCHs in the UL

integer set $\{0, \ldots, Q_k - 1\}$. The data symbols are multiplied by $w_{Q_k}^{(k)}$ before spreading. The values for the multiplier are given in Table D.1 in Appendix D.1.

7.4.3 Scrambling

After spreading, cell-specific scrambling is performed by applying a 16-chip long complex scrambling sequence $\underline{\nu} = (\underline{\nu}_1, \underline{\nu}_2, \ldots, \underline{\nu}_{16})$ whose elements are taken from $\{1, j, -1, -j\}$. The complex scrambling code $\underline{\nu}$ is generated from a binary code $\nu = (\nu_1, \nu_2, \ldots, \nu_{16})$ according to

$$\underline{\nu}_i = j^i \cdot \nu_i, \quad \nu_i \in \{1, -1\}; \quad i = 1, \ldots, 16. \tag{7.2}$$

Therefore, the elements of the scrambling code $\underline{\nu}$ alternate between real and imaginary. The binary code ν is taken out of 128 different codes specified in [124]. If the SF is not equal to 16, length matching is obtained by concatenating 16/SF spread data symbols.

7.4.4 Combination of Different Channels

Uplink. Two different UL DPCHs, which belong to the same CCTrCH, can be combined in one timeslot. After spreading and scrambling, the DPCHs are represented by complex-valued chip sequences. As depicted in Fig. 7.11, the amplitudes of the DPCHs are first adjusted according to the UL open loop power control (see section 7.7.1). Then each DPCH is weighted by a weight factor γ_i before the DPCHs are combined by complex addition. Subsequently, the combined DPCHs are weighted with a gain factor β_j depending on the actual TFCI. The gain factor accounts for the rate matching of the transport channels, which affects the transmit power required to achieve a certain E_b/N_0 [126]. The possible values of γ_i (which depend on the SF of the DPCH) and β_j are given in Appendix sections D.2 and D.3. They can change after

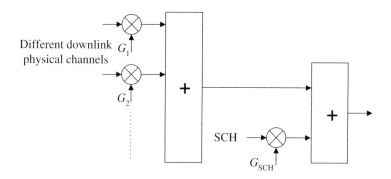

Fig. 7.12. Combination of different DPCHs in the DL for a timeslot in which the SCH is transmitted

each radio frame and are independent of any other type of power control. The schematic shown in Fig. 7.11 applies to each CCTrCH separately.

Downlink. Figure 7.12 shows the combination of different DPCHs in the DL. Each DPCH receives a weight factor G_i before the channels are complex added. The SCH is weighted separately with a factor G_{SCH} and is complex added to the combination of the DPCHs if it is transmitted in the timeslot.

7.4.5 Synchronization Codes

The primary synchronization code c_p is a 256-chip long complex code with identical real and imaginary parts, i.e., each element (chip) of the code words can only take values out of $\{+1 + j, -1 - j\}$. The code is generated in the same way as in the FDD mode from a so-called generalized hierarchical Golay sequence. This is described in Appendix D.4.

The 16 secondary synchronization codes $\{C_0, \ldots, C_{15}\}$ have a length of 256 chips and are complex valued with identical real and imaginary parts. They are constructed in a similar way to the FDD mode but with slightly different parameters as described in Appendix D.4. According to the two different cases for the allocation of the SCH (see section 7.2.5) the secondary synchronization codes are partitioned into two code sets for case 1 and four code sets for case 2. In Table 7.5 this partitioning is described. Three of the secondary synchronization codes are QPSK modulated and transmitted in parallel to the primary synchronization code. The QPSK modulation is carried out by multiplying each of the three secondary synchronization codes with either ± 1 or $\pm j$, e.g., for code group 9 in code set 1 (case 1), the following modulation is used: $jC_1, -jC_5, C_3$ in frame 1 (a frame with an odd frame number) and $jC_1, -jC_5, -C_3$ in frame 2 (a frame with an even frame number). The QPSK modulation of the secondary synchronization codes provides information about the code group of the BS, the position of the frame within

Table 7.5. Partitioning of the secondary synchronization codes

Case 1			Case 2		
Code set	Code group	Codes	Code set	Code group	Codes
1	0–15	C_1, C_3, C_5	1	0–7	C_1, C_3, C_5
2	16–31	C_{10}, C_{13}, C_{14}	2	8–15	C_{10}, C_{13}, C_{14}
			3	16–23	C_0, C_6, C_{12}
			4	24–31	C_4, C_8, C_{15}

an interleaving period of 20 ms, and the position of the SCH slot(s) within the frame. The modulated secondary SCH codes are constructed such in a way that a cyclic shift of less than 2 (Case 1) or 4 (Case 2) of any of the sequences is not identical to a cyclic shift from any other of the sequences. The tables with all possible modulated code combinations and their associated parameters can be found in [124].

7.4.6 Modulation

The QPSK modulation and the pulse shaping of the complex-valued chip sequence after spreading, scrambling, and combination of different DPCHs, if applicable, is shown in Fig. 7.13 and is the same as in the FDD mode. A detailed description is given in section 6.4.4

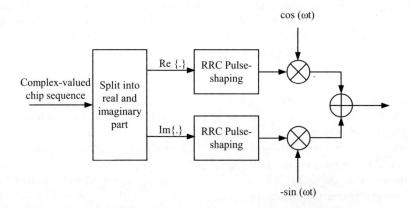

Fig. 7.13. QPSK modulation of the spread and scrambled complex-valued chip sequence

7.5 Spreading and Modulation for the 1.28 Mcps Option

7.5.1 Mapping of the Data Bits

Besides the mapping of the data bits to a QPSK constellation as in the TDD 3.84 Mcps option, 8PSK is possible with the 1.28 Mcps option. In this case, three consecutive bits are represented by one complex valued data symbol according to Table 7.6. Here, the superscript $i = 1, 2$ distinguishes between

Table 7.6. Mapping of the data bits b to 8PSK symbols $\underline{d}$

Consecutive binary bit pattern $b_{1,n}^{(k,i)} b_{2,n}^{(k,i)} b_{3,n}^{(k,i)}$	Complex symbol $\underline{d}_n^{(k,i)}$
000	$\cos(11\pi/8) + j\sin(11\pi/8)$
001	$\cos(9\pi/8) + j\sin(9\pi/8)$
010	$\cos(5\pi/8) + j\sin(5\pi/8)$
011	$\cos(7\pi/8) + j\sin(7\pi/8)$
100	$\cos(13\pi/8) + j\sin(13\pi/8)$
101	$\cos(15\pi/8) + j\sin(15\pi/8)$
110	$\cos(3\pi/8) + j\sin(3\pi/8)$
111	$\cos(\pi/8) + j\sin(\pi/8)$

the two data fields of the bursts, $k = 1, \ldots, K$ denotes the kth user, and the subscript $n = 1, \ldots, N_k$ identifies the nth data symbol out of a total of N_k symbols per data field for the user k. Figure 7.14 shows the constellation diagram of the 8PSK modulation format.

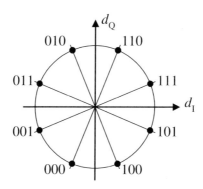

Fig. 7.14. Constellation diagram for 8PSK modulation

The channelization and scrambling operations and the combination of different channels in UL and DL are the same as with the TDD 3.84 Mcps option.

7.5.2 Synchronization Codes

Downlink Pilot Timeslot. One of the existing 32 different SYNC-DL sequences (specified in [124]), each with a length of 64 chips, is transmitted in the DwPTS after a guard period of 32 chips (see section 7.3.5). The SYNC-DL sequence number indicates the used code group in the cell. Each code group is associated with four different scrambling codes and four different basic midamble codes. The elements of the SYNC-DL codes alternate between real and imaginary and are not scrambled. The SYNC-DL sequences are QPSK modulated. The phases of four SYNCH-DL sequences (transmitted in four consecutive sub-frames) are used to signal the presence of the P-CCPCH in the following four sub-frames. If the presence of a P-CCPCH is indicated, the next following sub-frame is the first sub-frame of the interleaving period. Two different phase quadruples are possible:

- Sequence S1 $(135°, 45°, 225°, 135°)$ reports that there is a P-CCPCH in the next four sub-frames.
- Sequence S2 $(315°, 225°, 315°, 45°)$ reports that there is no P-CCPCH in the next four sub-frames.

Uplink Pilot Timeslot. There are 256 different SYNCH-UL sequences (= signatures) with a length of 128 chips, which are specified in [124]. A SYNCH-UL is transmitted in the UpPTS before a guard period of 32 chips. The elements of the SYNC-UL codes alternate between real and imaginary and are not scrambled.

7.6 Multiplexing, Channel Coding, and Interleaving

The steps to be performed during the coding, multiplexing, and interleaving procedure are the same as in the FDD mode [125]. The order in which the individual steps are performed in TDD UL and DL is the same as in the FDD UL shown in Fig. 6.20. The differences to FDD (as far as these steps are described in section 6.5) are the following:

Channel Coding. The coding schemes and rates for the different TrCHs are listed in Table 7.7 and Table 7.8 for the 3.84 Mcps and 1.28 Mcps options, respectively.

Second Interleaving. Two different schemes for second interleaving are possible. The second interleaving can be either applied jointly to all data bits transmitted during one frame or separately within each timeslot onto which the CCTrCH is mapped. Higher layers control the scheme selection.

Table 7.7. Channel coding schemes and coding rates for the 3.84 Mcps option of the TDD mode

Type of TrCH	Coding scheme	Coding rate
BCH	Convolutional coding	1/2
PCH	Convolutional coding	1/2
RACH	Convolutional coding	1/2
DCH, DSCH, FACH, USCH	Convolutional coding	1/3, 1/2
	Turbo coding	1/3
	No coding	

Table 7.8. Channel coding schemes and coding rates for the 1.28 Mcps option of the TDD mode

Type of TrCH	Coding scheme	Coding rate
BCH	Convolutional coding	1/2
PCH	Convolutional coding	1/2, 1/3
RACH	Convolutional coding	1/2
DCH, DSCH, FACH, USCH	Convolutional coding	1/3, 1/2
	Turbo coding	1/3
	No coding	

Sub-frame Segmentation for the 1.28 Mcps Option. In the 1.28 Mcps TDD option the required sub-frame segmentation is performed between the second interleaving and the physical channel mapping. The rate-matching operation guarantees an even number of bits that can be subdivided into two sub-frames.

7.7 Physical Layer Procedures

Further details on the physical layer procedures described in the following can be found in [126].

7.7.1 Transmit Power Control

The purpose of transmit power control (PC) applied in the TDD mode is to limit the interference level (intercell interference) within the system and to reduce the power consumption of the MS.

The same transmission power is used for all codes within one timeslot allocated to the same CCTrCH if they have the same spreading factor. Tables 7.9 and 7.10 describe the general transmit power control characteristics for the 3.84 Mcps and the 1.28 Mcps option, respectively.

Table 7.9. Transmit power control characteristics for the 3.84 Mcps option of the TDD mode

	Uplink	Downlink
Power control type	Open loop PC	SIR-based closed loop PC
Power control rate	Variable 1–7 slots delay (2 slot SCH) 1–14 slots delay (1 slot SCH)	Variable with rate depending on the slot allocation
Step size	–	1, 2, 3 dB
Remarks		Within one timeslot the powers of all active codes may be balanced to within a range of 20 dB

Table 7.10. Transmit power control characteristics for the 1.28 Mcps option of the TDD mode

	Uplink	Downlink
Power control type	SIR-based closed loop PC	SIR-based closed loop PC
Power control rate	Variable closed loop: $0 - 200$ cycles/s open loop: $200 - 3575$ μs delay	Variable closed loop: $0 - 200$ cycles/s
Step size	1, 2, 3 dB (closed loop)	1, 2, 3 dB (closed loop)
Remarks		Within one timeslot the powers of all active codes may be balanced to within a range of 20 dB

Uplink Power Control. Higher layers can set the total maximum allowed UL transmit power to a value lower than which the terminal is capable. If the sum of the transmit powers of all UL channels exceeds this value, the power of all channels is reduced by the same factor.

In the *3.84 Mcps option*, the UL transmit power of PRACH, DPCH, and USCH is controlled by an open loop power control mechanism after synchronization has been established. The transmit power P_{DPCH} of each UL DPCH in one CCTrCH and in one timeslot is computed according to [97]

$$P_{\text{DPCH}} = \alpha L_{\text{P–CCPCH}} + (1 - \alpha)L_0 + I_{\text{BTS}} + \text{SIR}_{\text{target}} + \text{Const}. \quad (7.3)$$

Here $L_{\text{P–CCPCH}}$ is the path loss in dB, L_0 is the long term average of the path loss in dB, I_{BTS} is the interference signal power level at the BS receiver in dBm (that is broadcast on the BCH) which can be timeslot specific, α (which can also be timeslot specific) is a weighting parameter representing the quality of path loss measurements and is computed in the MS, $\text{SIR}_{\text{target}}$ is the target SIR in dB for the CCTrCH, which is adjusted by a higher layer outer loop, and Const is a value set by higher layers. The transmit power of the PRACH is computed similarly as for the DPCH.

$$P_{\text{PRACH}} = L_{\text{P-CCPCH}} + I_{\text{BTS}} + \text{Const}.\tag{7.4}$$

The USCH transmit power is controlled by the same equation as the DPCH [97].

The *initial accuracy* of the power control setting in the *3.84 Mcps option* must be ± 9 dB under normal conditions and ± 12 dB under extreme conditions (see Appendix C.1). *Differential accuracy* defines the error in the MS transmitter power step due to changes of the parameters in (7.3). The *differential accuracy controlled input* has to be fulfilled if $\text{SIR}_{\text{target}}$ changes and $\alpha = 0$. $\text{SIR}_{\text{target}}$ is rounded to the closest integer dB value. The error must meet the criteria listed in Table 7.11. The *differential accuracy measured input* is specified for power changes due to a step change in the path loss $L_{\text{P-CCPCH}}$. The error for this case should not exceed the values specified in Table 7.11 plus the values specified for the measurement of the P-CCPCH RSCP (received signal code power) [131].

Table 7.11. Transmit power step tolerance for a step in the $\text{SIR}_{\text{target}}$

$\Delta\text{SIR}_{\text{target}}$ [dB]	Transmit power step tolerance [dB]
$\Delta\text{SIR}_{\text{target}} \leq 1$	± 0.5
$1 < \Delta\text{SIR}_{\text{target}} \leq 2$	± 1
$2 < \Delta\text{SIR}_{\text{target}} \leq 3$	± 1.5
$3 < \Delta\text{SIR}_{\text{target}} \leq 10$	± 2
$10 < \Delta\text{SIR}_{\text{target}} \leq 20$	± 4
$20 < \Delta\text{SIR}_{\text{target}} \leq 30$	± 6
$30 < \Delta\text{SIR}_{\text{target}}$	± 9 [1]

[1] Under normal conditions. For extreme conditions the value is ± 12 dB.

The initial UL transmit power for the DPCH and the PUSCH for the *TDD 1.28 Mcps option* is calculated as

$$P_{\text{DPCH}} = L_{\text{P-CCPCH}} + \text{SIR}_{\text{target}}.\tag{7.5}$$

After the MS starts to receive TPC bits on either DPCH or PUSCH it switches to closed loop power control on the respective channel with an overall dynamic range of 80 dB. The transmit power for UpPCH and PRACH are computed according to

$$P_{\text{UpPCH}} = L_{\text{P-CCPCH}} + \text{PRX}_{\text{UpPCHdes}} + i \cdot \text{Pwr}_{\text{ramp}}\tag{7.6}$$

and

$$P_{\text{PRACH}} = L_{\text{P-CCPCH}} + \text{PRX}_{\text{PRACHdes}} + i \cdot \text{Pwr}_{\text{ramp}}.\tag{7.7}$$

Here, $PRX_{UpPCHdes}$ is the desired UpPCH receive power, $PRX_{PRACHdes}$ is the desired PRACH receive power, i is the number of transmission attempts on the UpPCH, and Pwr_{ramp} is the "Power Ramp Step", by which the MS must increase its transmission power after every UpPCH transmission attempt without success. For further details see the description of the random access procedure for the TDD mode (7.7.8).

The requirements regarding the open loop power control for the *1.28 Mcps option* (which is applied only for the initial UL transmit power) are the same as with the 3.84 Mcps option. For the closed loop power control the step size Δ_{TPC} is specified to be 1, 2, or 3 dB. Table 7.12 lists the necessary accuracy for the different step sizes. The power steps are defined as the

Table 7.12. Transmit power step size accuracy for the closed loop UL power control for the 1.28 Mcps option of the TDD mode

TPC_cmd	1 dB	2 dB	3 dB
Tolerance	±0.5 dB	±1.0 dB	±1.5 dB

relative power difference between the average power of the original timeslot and the average power of the target timeslot, not including the transient duration. If a sequence of 10 identical TCP commands is transmitted to the BS, the the difference between the original transmit power and the transmit power after the 10 equal steps must be within tighter limits than 10 times the limits specified in Table 7.12 (see [129]).

Downlink Power Control. The initial transmission power of the DL DPCH is set by the UTRAN. After the initial transmission an SIR-based inner loop power control is applied. A higher layer outer loop adjusts the target SIR in order to achieve the quality target (which is the transport channel BLER (Block Error Rate) value for each transport channel set by the UTRAN). The MS sets the SIR target when the physical channel has been set up or reconfigured. It is not allowed to increase the SIR target value before the power control has converged on the current value. The measurement of the received SIR is carried out periodically at the MS. If the measured SIR is higher than the target SIR value, the transmitted TPC command is "down", otherwise it is "up". In response to the TPC commands the BS transmit power is raised or reduced by one step as long as the limits set by higher layers are not exceeded. The average power over one slot of the complex QPSK symbols of a single DPCH before spreading is defined as transmit power. The power control step size tolerance is specified in Table 7.13 If the total transmit power of all DL DPCHs exceeds the maximum allowed transmit power, the power of all DPCHs is reduced by the same amount to fulfill the requirement.

Table 7.13. DL transmit power control step size tolerance

TPC_cmd	1 dB	2 dB	3 dB
Tolerance	±0.5 dB	±0.75 dB	±1 dB

7.7.2 Timing Advance (3.84 Mcps Option)

The timing of the slots transmitted by each MS has to be adjusted to guarantee that the signals of all MSs arrive at the BS with the correct timing regardless of the distance of the MSs from the BS. The parameter which controls the timing adjustment is called *timing advance*. Its initial value is determined by the UTRAN from the timing measurement of the PRACH (see section 7.2.2). The actual timing advance value is an integer multiple of 4 chips which is nearest to the required timing advance. The integer value is between 0 and 63 and is encoded in a 6-bit number. The UTRAN continuously measures the timing of the MSs transmission and returns the necessary timing advance value to the MS by means of higher-layer messages. The MS adjusts its transmission timing according to the timing advance command. This adjustment is done at the frame number specified by higher-layer signaling. In the case of a handover, the timing advance (TA) in the new cell has to be adjusted by the relative timing difference Δt between the new and the old cell: $TA_{new} = TA_{old} + 2\Delta t$.

If UL synchronization is used, the timing advance has sub-chip step size and high accuracy to enable synchronous CDMA in the UL. In this case, the required timing advance is represented as a multiple of a 1/4 chip. The support of UL synchronization is optional for the MS.

7.7.3 UL Synchronization (1.28 Mcps Option)

UL synchronization, which can start only after DL synchronization has been established, is performed during the random access procedure. The first UL transmission is performed in the special UpPTS to reduce interference in the traffic timeslots. The UpPTS features an extended guard period of 32 chips to account for the unknown distance to the BS. The MS adjusts the transmission time $T_{Tx,UpPCH}$ of the UpPCH (sent in the UpPTS) according to

$$T_{Tx,UpPCH} = T_{Rx,DwPCH} - 2\delta t_p + 12 \cdot 16 \cdot T_{chip}. \tag{7.8}$$

Here, $T_{Rx,DwPCH}$ is the received starting time of the DwPCH with respect to the timing of the MS and δt_p is the timing advance of the UpPCH in multiples of $1/8 T_{chip}$ which is estimated in the MS.

The BS evaluates the timing of the detected SYNC-UL sequence (see section 7.5.2) and provides the timing deviation information (UpPCH_POS) to the MS via the FPACH. The timing deviation is represented as an 11 bit number (0–2047), being a multiple of $1/8 T_{chip}$, which is nearest to the received position of the UpPCH. The MS uses the $UpPCH_{POS}$ value to adjust the timing advance of the PRACH accordingly.

After the UL synchronization has been established in the MS, a continuous closed loop UL synchronization control procedure starts for the transmission of the PUSCH and the DPCH. The BS can use each UL burst to measure the timing of the MS and send the necessary SS commands in each sub-frame to the MS. If necessary, the MS adjusts its timing accordingly in steps of $\pm k/8T_{chip}$ each M sub-frames. The default values of M (1–8) and k (1–8) are broadcast on the BCH. The values of M and k can be adjusted during a call setup or readjusted during the call.

7.7.4 Cell Search for the 3.84 Mcps Option

The same three steps as in the FDD mode are performed during the initial cell search but with slightly different details. The SCH (see section 7.2.5) is used to acquire all necessary information in the cell search procedure.

Primary Synchronization Code Acquisition. The primary synchronization code transmitted on the SCH can be used to find a cell with a single matched filter, because this code is unique throughout the system. In case 1 (see section 7.2.5) the SCH is received in every 15th slot, while in case 2 the second SCH slot is received in the 7th or 8th slot after the first SCH slot.

Codegroup Identification and Slot Synchronization. In the second step, the QPSK-modulated secondary synchronization codes $\{C_0, \ldots, C_{15}\}$ are used to identify the code group of the cell. The codes and their modulation are found by correlating the possible 16 secondary synchronization codes with the received signal in the slots found in step 1. The modulated codes are constructed in a way that their cyclic-shifts are unique, i.e., a non-zero cyclic shift less than 2 (Case 1) and 4 (Case 2) of any of the sequences is not equivalent to some cyclic shift of any other of the sequences. Also, a non-zero cyclic shift less than 2 (Case 1) and 4 (Case 2) of any of the sequences is not equivalent to itself with any other cyclic shift less than 8. The 32 code groups are determined by the combinations of modulated secondary synchronization codes specified in [124] for the two different cases described in section 7.2.5. The primary synchronization code provides the phase reference for the coherent detection of the three secondary synchronization codes. Each code group is associated with a timing offset and with four specific cell parameters each of which is linked to a particular short and long basic midamble code. This scheme enables the MS to find the frame border by using the position of the SCH and the knowledge of t_{offset}. The modulation of the secondary synchronization codes also indicates the position of the SCH slot within a two-frame period.

Scrambling Code Identification and Frame Synchronization. In the third and last step of the initial cell search procedure, the basic midamble code and the associated scrambling code is found by the MS. They are identified by means of correlation over the P-CCPCH with all four midambles of the already known code group. This is possible since the P-CCPCH, which contains the BCH, is transmitted using the first channelization code $c_{Q=16}^{k=1}$ and the first midamble code. With the knowledge of the scrambling code, the MS can read system- and cell-specific information from the BCH and acquire frame synchronization.

7.7.5 Cell Search for the 1.28 Mcps Option

The initial cell search in the 1.28 Mcps option is performed in four steps:

Search for DwPTS. Matched filters can be used in the MS for detecting the SYNC-DL in order to acquire DwPTS synchronization to a cell. The MS has to identify one out of 32 possible SYNC-DL sequences.

Scrambling and Basic Midamble Code Identification. In the second step of the cell search procedure, the MS determines the basic midamble code from the SYNC-DL sequence number. Each SYNC-DL sequence number is associated with four different basic midamble codes. The MS can determine the used basic midamble code using a trial and error technique. This basic midamble code is used throughout the frame. Each basic midamble code is associated with a scrambling code. Therefore, the scrambling code is also known when the basic midamble code is identified.

Control Multi-frame Synchronization. During the third step of the initial cell search procedure, the MS detects the sequence of four consecutive phases of the SYNC-DL on the DwPTS, which are used to signal the presence of the P-CCPCH in the following four sub-frames (see section 7.5.2).

Read the BCH. The initial cell search is finished if the cell specific control information is read from the BCH.

7.7.6 Discontinuous Transmission

If the bit rate of a CCTrCH, mapped to dedicated and shared physical channels (PUSCH, PDSCH, UL DPCH and DL DPCH), differs from the total channel bit rate, rate matching is applied. Rate matching fills resource units completely with data. In the case that after rate matching no data at all are to be transmitted in a resource unit, the complete resource unit is discarded from transmission which yields DTX.

If no transport blocks provided by higher layers have to be transmitted, a so-called "Special Burst" is sent in the first allocated frame of the transmission pause. If there is a consecutive period of Special Burst Period (SBP) frames without transmission of transport blocks, another special burst is generated and transmitted at the next possible frame. This pattern is continued until transport blocks are provided for the CCTrCH by the higher layers. The

SBP is independently specified for uplink and for downlink and is provided by higher layer signaling. The default value for both UL and DL SBP is 8.

The special burst has the same slot format as the normal burst used for data transmission. It contains an arbitrary bit pattern, TFCI, and TPC bits if inner loop PC is applied. The TFCI of the special burst is filled with "0" bits. The special burst is transmitted with the same power as that of the substituted physical channel of the CCTrCH carrying the TFCI.

In the 1.28 Mcps option, DTX is applied in the same way as in the 3.84 Mcps option with the special burst being transmitted in both consecutive sub-frames.

7.7.7 Downlink Transmit Diversity

The support for downlink transmit diversity for DPCH, P-CCPCH, PDSCH, and SCH in the 3.84 Mcps option and for DPCH, P-CCPCH, and DwPTS in the 1.28 Mcps option is optional for the core network and mandatory for the MS. For the DL DPCH in both options and the PDSCH, closed loop transmit diversity is realized by passing the chip stream in the BS to two distinct transmit branches (pulse shaping, D/A conversion, translation to RF, antenna). Each path is multiplied with an antenna-specific complex weight factor (w_1 and w_2), which is calculated for each slot and user. The weight factors are determined by the UTRAN. Time switched transmit diversity can be employed for the SCH (3.84 Mcps) and the DwPTS (1.28 Mcps). In this case the SCH/DwPTS is transmitted from antenna 1 and antenna 2, alternatively. In the 1.28 Mcps option, TSTD can also be applied to the DL DPCH. In this case not all DPCHs in a sub-frame need to be transmitted on the same antenna and not all DPCHs within a sub-frame have to use TSTD.

The block space time transmit diversity, which can be applied to the P-CCPCH in both options, is shown in Fig. 7.15. The channel coding, rate matching, interleaving, and bit-to-symbol mapping is performed as in the non-diversity mode. The block STTD encoding is separately performed for each of the two data fields in a burst. Each data field at the encoder input generates two data fields at the output, corresponding to each of the diversity antennas. Spreading and scrambling is performed after the STTD encoder.

7.7.8 Random Access Procedure

If a higher layer requests the transmission of a message on the RACH, the physical random access procedure is invoked.

Random Access Procedure for the 3.84 Mcps Option. Before any transmission on the RACH takes place the MS has to decode information from the BCH. This necessary *a priori* information consists of: the available PRACH sub-channels and associated Access Service Classes (ASC), the timeslot, spreading factor, channelization code, midamble, repetition period,

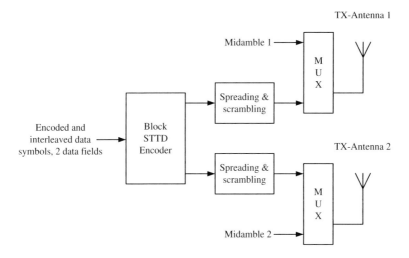

Fig. 7.15. Block STTD schematic for the P-CCPCH

and offset for each PRACH sub-channel, the set of transport format parameters, and the parameters of the common physical channel uplink outer loop power control. The following steps are performed during the physical random access procedure:

- Layer 1 receives the transport format for the PRACH message, the ASC, and the data to be transmitted from the MAC layer.
- A spreading code is randomly selected (with equal probability). The midamble is derived from the selected spreading code.
- The MS randomly (with equal probability) selects a PRACH sub-channel.
- The data are transmitted with no timing advance using a transmission power level according to the specification for common physical channels in the UL.

Higher layers (MAC and RRC) control the parameters of random access procedure like, e.g., backoff algorithm and timer handling for re-transmission in case of failed RACH transmission due to collision.

Random Access Procedure for the 1.28 Mcps Option. The MS reads the cell broadcast information (e.g., codes, spreading factor, midambles, and timeslots for PRACH, FPACH, and S-CCPCH, which carries the FACH) and derives the code set of the 8 SYNC-UL codes (= signatures) assigned to the UpPCH for random access from the used SYNC-DL code on the DwPCH. Before the random access procedure starts, the following information must be passed from the MAC layer to layer 1:

- The association between signatures (SYNCH-UL codes) and FPACHs, between FPACHs and PRACHs, and PRACHs and CCPCHs, and the parameter values for each of these physical channels.

- The length L_i of a RACH message associated to FPACH$_i$, which can be configured to be either 1, 2, or 4 sub-frames corresponding to a length in time of either 5, 10, or 20 ms.
- The available UpPCH sub-channels for each ASC.
- The set of Transport Format parameters for the PRACH message.
- The maximum number of transmissions on the UpPCH.
- The maximum number of sub-frames to wait for the network acknowledgement to a sent signature. The maximum value supported by layer 1 is 4 sub-frames.
- The initial signature power.

The following steps are performed during the physical random access procedure:

- At the initiation of the random access procedure layer 1 receives the transport format for the PRACH message, the ACS, and the data to be transmitted from the MAC layer.
- The signature re-transmission counter and the signature transmit power are set.
- The MS randomly (with equal probability) selects an UpPCH sub-channel and transmits the signature (SYNCH-UL code).
- The BS measures the timing deviation with respect to the reference time of the received first path from the UpPCH. It sends the FPACH burst on the FPACH$_i$ associated to the UpPCH. The FPACH burst contains the acknowledgement of the detected signature and timing and power level information for the RACH message.
- The MS detects the FPACH$_i$ to receive the acknowledgement from the UTRAN.
- If no valid answer is detected in due time by the MS the signature re-transmission counter is decreased by one and a signature is again transmitted on the UpPCH if the counter is greater than 0. Otherwise a random access failure is reported to the MAC layer.
- If a valid answer is detected by the MS the RACH message is sent on the PRACH 2 sub-frames after the sub-frame in which the acknowledgement was received.

If a collision is very likely or a bad propagation environment appears, the BS will not transmit the FPACH or cannot receive the SYNC-UL. In this case, the MS does not receive the acknowledgement from the UTRAN at all and has to adjust its transmission time and power level. Owing to this two-step mechanism for the random access procedure, collisions happen almost exclusively on the UpPCH while the RACH is basically free from collisions and can be transmitted together with conventional data channels in the same timeslots.

7.7.9 DSCH Procedure

Three methods are available to tell the MS that there are data to decode on the DSCH. These are: (1) The TFCI field of the associated channel or PDSCH. (2) The user specific midamble of the DSCH. (3) Higher layer signaling. The described DSCH procedure is applied if either method 1 or 2 is used.

DSCH Procedure with TFCI Indication. If the transmission of data on the DSCH for the MS is indicated via the TFCI, two cases can be distinguished:

- The TFCI is located on the PDSCH itself in the case of a stand-alone PDSCH. The MS decodes the data parts of the PDSCH only if the decoded TFCI field corresponds to the TFC part of the TFCS (TFC set) given to the MS by higher layers.
- If the TFCI is located on the DCH, the MS must start decoding the PDSCH two frames after the DCH frame that contained the TFCI indicating the use of the DSCH. If the TFCI is transmitted over several frames, the PDSCH must be decoded starting two frames after the frame that contained the last part of the TFCI.

DSCH Procedure with Midamble Indication. If midamble indication for the use of the DSCH is used, the MS has to test the received midamble if it corresponds to the midamble that was assigned to indicate DSCH reception. If this is the case, the data of the PDSCH are decoded according to the TFCI, which can be either located on the DCH or the PDSCH itself.

7.7.10 Node B Synchronization Procedure over the Air

In this optional procedure the cell sync bursts (see section 7.2.8) can be used to establish and maintain BS synchronization [127]. This type of synchronization is the achievement of a common timing reference among different BSs. Although not required, such a common timing reference can be useful in order to minimize the buffering time and the transmission delay for the DL transmission on the air interface.

The cell synch bursts are transmitted in timeslots normally assigned to the PRACH. They allow the measurement of timing offsets between the cells, which are reported to the UTRAN. The UTRAN in turn can adjust the timing of the BSs. Cell synchronization is acquired and maintained in three different phases:

- During the frequency acquisition phase the cells to be synchronized are listening to the continuous transmission of cell synch bursts of a cell that serves as the master time reference. All cells lock their frequency reference to the master cell. During this phase no traffic is supported.

- In the initial synchronization phase all cells transmit cell sync bursts with the same code and code offset according to the higher layer commands. All cells also evaluate the transmissions from the other cells and report the timing and received SIR of successfully detected cell sync bursts back to the UTRAN. The network uses these measurements to adjust the transmission timing of each cell. During this phase no traffic is supported.
- In the steady-state phase cell synch bursts are transmitted according to higher layer commands to maintain the synchronization. The transmission is scheduled in such a way that it does not disturb the other data traffic.

7.7.11 Idle Periods in DL (IPDL)

Idle periods can be created in the DL to support time difference measurements for location services [128]. During these idle periods, which have a duration of one slot, only the SCH is transmitted from a BS which improves the reception of signals from neighboring cells at the MS. Idle periods are generated according to higher layer commands in either a continuous mode or in a burst mode. In the continuous mode idle periods occur steadily. In the burst mode several idle periods, which are sufficient to perform all necessary measurements, are grouped together in a so-called burst. No idle periods are generated between bursts. Further details of the IPDL can be found in [126].

7.8 RF Characteristics: Transmitter Requirements and Receiver Test Cases for the TDD Mode

In this section we will describe the RF characteristics and test cases to be fulfilled by the MSs [129] and BSs [130] in TDD mode. Most of the parameters in the following subsections are referred to the 12.2 kbps UL/DL reference measurement channels with UL power control ON and DL power control OFF. The power values are averaged over at least one timeslot excluding any transient periods and measured with the root-raised cosine filter used for transmit pulse shaping (section 6.4.4). Only differences to the FDD mode will be described in the following. Therefore, if a specification described in the FDD chapter is not included in this subsection, it is the same in TDD mode as in the FDD mode. If a description is given only for the 1.28 Mcps option, the parameters for the 3.84 Mcps option are identical to the FDD mode.

7.8.1 Frequency Bands and Channel Arrangement in TDD Mode

As specified in Table 7.1, the so-called unpaired frequency bands for the UTRA TDD mode are

- 1900 − 1920 MHz and 2010 − 2025 MHz for UL and DL transmission (Band A).

In ITU region 2 (which comprises the countries of North and South America) these bands are at

- 1850 − 1910 MHz
- 1930 − 1990 MHz (Band B) and
- 1910 − 1930 MHz (Band C), all for UL and DL transmission.

The channel spacing is 5 MHz in the 3.84 Mcps option and 1.6 MHz in the 1.28 Mcps option. This spacing can be adjusted to optimize the system performance for a given deployment scenario. The channel center frequency raster to be used for such an adjustment is 200 kHz wide, i.e., the channel center frequency must be an integer multiple of 200 kHz. Within the UTRA framework a carrier frequency is uniquely defined by its UARFCN (see section 6.8.1).

7.8.2 MS Transmitter Requirements for the TDD Mode

The MS transmitter requirements for the TDD mode described in the following have to be fulfilled for the 12.2 kbps UL reference measurement channel whose parameters are listed in Table 7.14 for the 3.84 Mcps option and in Table 7.15 for the 1.28 Mcps option.

Table 7.14. Physical parameters for the 12.2 kbps UL reference measurement channel for the 3.84 Mcps TDD option

Information bit rate	12.2 kbps
Power control	On
TFCI	On
SF	8
Coding	Convolutional coding
Code rate	1/3
Midamble	512 bits

Table 7.15. Physical parameters for the 12.2 kbps UL reference measurement channel for the 1.28 Mcps TDD option

Information bit rate	12.2 kbps
Power control	On
TFCI	On
SF	8
Coding	Convolutional coding
Code rate	1/3
Midamble	144 bits

Table 7.16. Power classes for MSs in TDD mode

Power class	Maximum output power	Tolerance
1	+30 dBm	+1/ − 3 dBm
2	+24 dBm	+1/ − 3 dBm
3	+21 dBm	±2 dBm
4	+10 dBm	±4 dBm

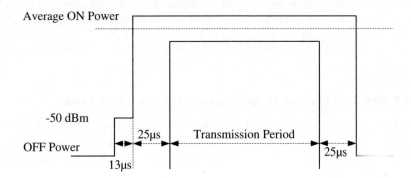

Fig. 7.16. Mask for power ramping between ON and OFF state for the 3.84 Mcps TDD mode

Transmit Power in the 3.84 Mcps Option. All RF parameters are defined for the MS in the release 4 specifications only for power classes 2 and 3 (see Table 7.16). The maximum output power is the average power of the transmit timeslots at the maximum power control setting. The averaging has to be performed over the useful part of the transmit timeslots. As in FDD mode, the power is measured with a root-raised cosine filter with a roll-off factor $\alpha = 0.22$ and a bandwidth equal to the chip rate. For multi-code operation the maximum output power has to be reduced by the difference of the peak-to-average power ratio between multi-code and single-code transmission. For MSs which use directive antennas for transmission, a class dependent limit will be set for the maximum EIRP (Equivalent Isotropic Radiated Power). Since the TDD mode is primarily intended for use in pico-cells and indoor environments, the possibility of a power class 4 with a maximum output power of only 10 dBm is given.

The minimum required transmit output power must be higher than −44 dBm. In the transmit OFF power state (the MS transmitter is turned off) the maximum transmit output power must be below −65 dBm. During the transitions between ON and OFF state the ramping time mask shown in Fig. 7.16 applies. The transmission period contains the burst without guard period for a single transmission slot. In case of consecutive transmission slots

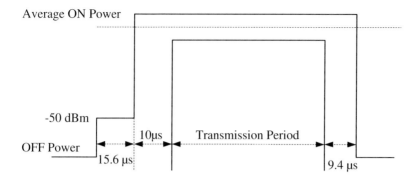

Fig. 7.17. Mask for power ramping between ON and OFF state for the 1.28 Mcps TDD mode

the transmission period starts at the beginning of the burst in the first transmission slot and stops before the guard period at the end of the last burst.

Transmit Power in the 1.28 Mcps Option. Differences with respect to the 3.84 Mcps option occur with the power control specifications (see section 7.7.1), the minimum output power, and the power ramping. The minimum required transmit output power must be higher than -49 dBm. During the transitions between ON and OFF state the ramping time mask shown in Fig. 7.17 applies.

RF Emissions. The occupied bandwidth in the 1.28 Mcps TDD option, which is defined so as to contain 99% of the total integrated power of the transmitted spectrum centered on the assigned channel frequency, must be less than 1.6 MHz for the chip rate of 1.28 Mcps. The spectrum emission mask (applicable to frequencies between 0.8 and 4.0 MHz away from the carrier frequency) for the 1.28 Mcps option is shown in Fig. 7.18. The spectrum emission mask for spurious emissions (due to harmonics emission, parasitic emission, intermodulation products, and frequency conversion products) in the 1.28 Mcps TDD option is plotted in Fig. 7.19 and applies only for frequencies more than 4 MHz away from the carrier frequency.

The ACLR requirements in the TDD mode, 3.84 Mcps option, are the same as in the FDD mode. For the 1.28 Mcps option, the adjacent channels are spaced ±1.6 MHz and ±3.2 MHz away from the transmit channel, and the ACLR must be better than 33 dB (±1.6 MHz) and 43 dB (±3.2 MHz) if the absolute adjacent channel power is higher than -55 dBm/1.28 MHz.

Transmit Modulation. The pulse shaping filter in the transmitter is the same root-raised cosine filter as used for the FDD mode, but for the 1.28 MHz option the chip duration is $T_{\text{chip}} = 0.78125\,\mu\text{s}$. The PCDE must stay below -21 dB at a SF of 16 for a measurement interval of one timeslot. The 12.2 kbps UL multi-code reference measurement channel used for the PCDE measurement is the same as the 12.2 kbps reference measurement channel described

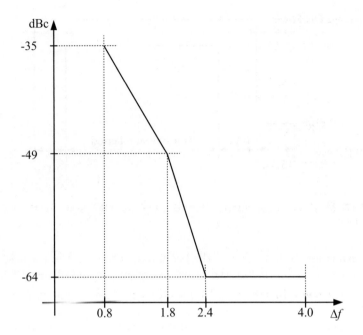

Fig. 7.18. Spectrum emission mask for out-of-band emissions for the 1.28 Mcps TDD option measured with a bandwidth of 30 kHz

at the beginning of this section with the only difference that the used SF is 16 instead of 8 which results in two code channels to be transmitted in parallel.

7.8.3 BS Transmitter Requirements for the TDD Mode

Unless otherwise noted, Table 7.17 lists the conditions for the test of the subsequently described requirements. The transmitted data on the DPDCHs must be sufficient random to emulated real data.

Transmit Power. The dynamic range of the DL power control must be 30 dB and the minimum controlled transmit power (when the power control setting is set to a minimum value) must be −30 dBm. In the transmit OFF power state (the BS transmitter is turned off) the maximum transmit output power must stay below −79 dBm in the 3.84 Mcps option and below −82 dBm in the 1.28 Mcps option. During the transitions between ON and OFF state the ramping time mask shown in Fig. 7.20 applies. Figure 7.21 shows the mask to be fulfilled for the 1.28 Mcps option.

Out-of-band Emissions. The spectrum emission masks for the 3.84 Mcps option is identical to the FDD mode. The spectrum emission masks for the 1.28 Mcps option are given in the Tables 7.18, 7.19, 7.20, and 7.21 and are

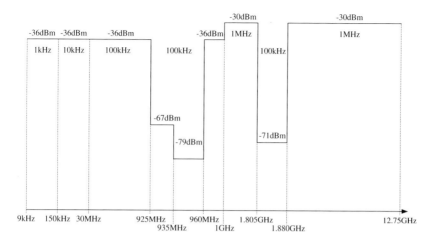

Fig. 7.19. Spectrum emission mask for spurious emissions for the 1.28 Mcps TDD option. For each frequency band the measurement bandwidth is given separately

Table 7.17. Parameters for the testing of the BS transmitter for the TDD mode

Parameter	Value
TDD duty cycle	Transmit in the even and receive in the odd numbered timeslots
BS transmit power	Rated output power of the BS
Number of active DPCH's in a timeslot	9 (3.84 Mcps option) 8 (1.28 Mcps option)
Power of each DPCH	1/9 (3.84 Mcps option) 1/8 (1.28 Mcps option) of the BS transmit power

summarized in Fig. 7.22. They cover the out-of-band emissions resulting from modulation and nonlinearities in the transmitter. The parameter Δf is the frequency separation between the carrier frequency and the $-3\,\mathrm{dB}$ point of the measuring filter closest to the carrier frequency. The separation between the carrier frequency and the center frequency of the measurement filter is denoted by f_{off} and $f_{\mathrm{off,Max}}$ is either 4 MHz or the offset to the UMTS transmit band edge, whichever is higher.

The ACLR (defined in section 6.8.2) must be better than 45 dB for the neighboring channels spaced ±5 MHz away from the transmit channel and better than 55 dB for the channels spaced ±10 MHz away in the 3.84 Mcps option. If the BS is operated in the proximity to another TDD BS or a FDD BS operating on the first or second adjacent channel center frequency, the ACLR must be better than 70 dB for both ±5 and ±10 MHz channel offset. This requirement is based on the assumption that the coupling loss between

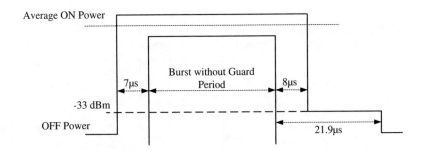

Fig. 7.20. Mask for power ramping between ON and OFF state for BSs for the 3.84 Mcps TDD option

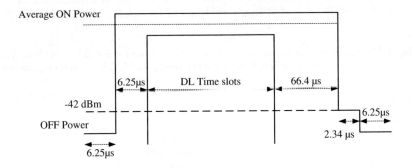

Fig. 7.21. Mask for power ramping between ON and OFF state for BSs for the 1.28 Mcps TDD option

Table 7.18. Spectrum emission mask values for a BS maximum output power of $P \geq 43\,\text{dBm}$ for the 1.28 Mcps TDD option

Δf	f_{off}	Maximum emission level	Measurement bandwidth
MHz	MHz	dBm	kHz
$0.8 \leq \Delta f < 1.0$	$0.815 \leq f_{\text{off}} < 1.015$	-14	30
$1.0 \leq \Delta f < 1.8$	$1.015 \leq f_{\text{off}} < 1.815$	$-14-15(f_{\text{off}}-1.015)$	30
	$1.815 \leq f_{\text{off}} < 2.3$	-28 .	30
$1.8 \leq \Delta f$	$2.3 \leq f_{\text{off}} < f_{\text{off,Max}}$	-13	1000

the base stations is at least 84 dB [132]. If a TDD BS or a FDD BS operating on an adjacent frequency channel is located at the same site, the ACLR must be better than 80 dB for both ±5 and ±10 MHz channel offset.

For the 1.28 Mcps option, the basic ALCR limits are 40 dB and 50 dB for neighboring channels at ±1.6 MHz and ±3.2 MHz away from the transmit channel, respectively. For the operation of the BS in the proximity of another TDD BS or a FDD BS and in the case that both BSs operate in adjacent

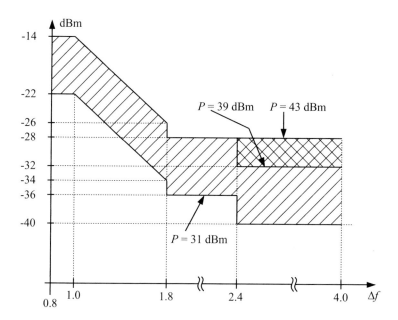

Fig. 7.22. Spectrum emission mask for the out-of-band emissions for the 1.28 Mcps TDD option measured with a bandwidth of 30 kHz

Table 7.19. Spectrum emission mask values for a BS maximum output power of $39 \leq P < 43$ dBm for the 1.28 Mcps TDD option

Δf	f_{off}	Maximum emission level	Measurement bandwidth
MHz	MHz	dBm	kHz
$0.8 \leq \Delta f < 1.0$	$0.815 \leq f_{\text{off}} < 1.015$	-14	30
$1.0 \leq \Delta f < 1.8$	$1.015 \leq f_{\text{off}} < 1.815$	$-14 - 15(f_{\text{off}} - 1.015)$	30
$1.8 \leq \Delta f < 2.4$	$1.815 \leq f_{\text{off}} < 2.415$	-28	30
	$2.415 \leq f_{\text{off}} < 2.9$	$P - 71$	30
$2.4 \leq \Delta f$	$2.9 \leq f_{\text{off}} < f_{\text{off,Max}}$	$P - 56$	1000

frequency bands, the maximum level of interference power must stay below -36 dBm. In the case of multiple antennas, the interference powers at all antenna connectors must be summed up. This applies for the co-existence of non-frame and non-switching point synchronized systems. If a TDD BS or a FDD BS operating on an adjacent frequency channel is located at the same site, the maximum level of interference power must stay below -76 dBm.

Spurious Emissions. For category B, (see subsection 6.8.3) the spurious emission limits are listed in Table 7.22 for the 3.84 Mcps option and in Table 7.23 for the 1.28 Mcps option. The frequencies F_{c1} and F_{c2} are the center

Table 7.20. Spectrum emission mask values for a BS maximum output power of $31 \leq P < 39$ dBm for the 1.28 Mcps TDD option

Δf	f_{off}	Maximum emission level	Measurement bandwidth
MHz	MHz	dBm	kHz
$0.8 \leq \Delta f < 1.0$	$0.815 \leq f_{\text{off}} < 1.015$	$P - 53$	30
$1.0 \leq \Delta f < 1.8$	$1.015 \leq f_{\text{off}} < 1.815$	$P - 53 - 15(f_{\text{off}} - 1.015)$	30
$1.8 \leq \Delta f < 2.4$	$1.815 \leq f_{\text{off}} < 2.415$	$P - 67$	30
	$2.415 \leq f_{\text{off}} < 2.9$	$P - 71$	30
$2.4 \leq \Delta f$	$2.9 \leq f_{\text{off}} < f_{\text{off,Max}}$	$P - 56$	1000

Table 7.21. Spectrum emission mask values for a BS maximum output power of $P < 31$ dBm for the 1.28 Mcps TDD option

Δf	f_{off}	Maximum emission level	Measurement bandwidth
MHz	MHz	dBm	kHz
$0.8 \leq \Delta f < 1.0$	$0.815 \leq f_{\text{off}} < 1.015$	-22	30
$1.0 \leq \Delta f < 1.8$	$1.015 \leq f_{\text{off}} < 1.815$	$-22 - 15(f_{\text{off}} - 1.015)$	30
$1.8 \leq \Delta f < 2.4$	$1.815 \leq f_{\text{off}} < 2.415$	-36	30
	$2.415 \leq f_{\text{off}} < 2.9$	-40	30
$2.4 \leq \Delta f$	$2.9 \leq f_{\text{off}} < f_{\text{off,Max}}$	-25	1000

frequencies of the first and last carrier transmitted by the BS, respectively. The frequencies F_l and F_u are the lower and the upper frequency of the TDD frequency band, respectively.

The spurious emission limits for the coexistence with other mobile communication standards are the same as in the FDD mode. The limits for the coexistence with the FDD mode are given in Table 7.24.

Transmit Modulation. The EVM must stay below 12.5 % and the Peak Code Domain Error limit is -28 dB for a spreading factor of 16. The EVM test must be performed with just one DPCH active.

7.8.4 MS Receiver Test Cases for the TDD Mode

Table 7.25 lists common terms used in the UMTS TDD specifications [129, 133] for defining the MS receiver requirements to be fulfilled. Unless otherwise stated, all parameters are specified at the antenna connector of the MS.

All requirements for the receiver test cases for TDD MSs have to be fulfilled in case of the transmission of the 12.2 kbps DL reference measurement channel with the parameters listed in Table 7.26 for the 3.84 Mcps option and in Table 7.27 for the 1.28 Mcps option. The coded bit error rate under the conditions stated for each receiver test case must not exceed 10^{-3}.

Table 7.22. Spurious emission limits for BSs for the 3.84 Mcps TDD option according to category B

Band	Maximum level	Measurement bandwidth
9 kHz − 150 kHz	−36 dBm	1 kHz
150 kHz − 30 MHz	−36 dBm	10 kHz
30 MHz − 1 GHz	−36 dBm	100 kHz
1 GHz − $(F_{c1} - 60\,\text{MHz})$ or $(F_1 - 10\,\text{MHz})$ whichever is higher	−30 dBm	1 MHz
$(F_{c1} - 60\,\text{MHz})$ or $(F_1 - 10\,\text{MHz})$ whichever is higher − $(F_{c1} - 50\,\text{MHz})$ or $(F_1 - 10\,\text{MHz})$ whichever is higher	−25 dBm	1 MHz
$(F_{c1} - 50\,\text{MHz})$ or $(F_1 - 10\,\text{MHz})$ whichever is higher − $(F_{c2} + 50\,\text{MHz})$ or $(F_u + 10\,\text{MHz})$ whichever is lower	−15 dBm	1 MHz
$(F_{c2} + 50\,\text{MHz})$ or $(F_u + 10\,\text{MHz})$ whichever is lower − $(F_{c2} + 60\,\text{MHz})$ or $(F_u + 10\,\text{MHz})$ whichever is lower	−25 dBm	1 MHz
$(F_{c2} + 60\,\text{MHz})$ or $(F_u + 10\,\text{MHz})$ whichever is lower − 12.75 GHz	−30 dBm	1 MHz

The following steps are performed to achieve the DPCH data rate of 47.2 kbps for the 3.84 *Mcps option* from the information bit rate of 12.2 kbps

Table 7.23. Spurious emission limits for BSs for the 1.28 Mcps TDD option according to category B

Band	Maximum level	Measurement bandwidth
9 kHz − 150 kHz	−36 dBm	1 kHz
150 kHz − 30 MHz	−36 dBm	10 kHz
30 MHz − 1 GHz	−36 dBm	100 kHz
1 GHz − $(F_{c1} - 19.2\,\text{MHz})$ or $(F_1 - 3.2\,\text{MHz})$ whichever is higher	−30 dBm	1 MHz
$(F_{c1} - 19.2\,\text{MHz})$ or $(F_1 - 3.2\,\text{MHz})$ whichever is higher − $(F_{c1} - 16\,\text{MHz})$ or $(F_1 - 3.2\,\text{MHz})$ whichever is higher	−25 dBm	1 MHz
$(F_{c1} - 16\,\text{MHz})$ or $(F_1 - 3.2\,\text{MHz})$ whichever is higher − $(F_{c2} + 16\,\text{MHz})$ or $(F_u + 3.2\,\text{MHz})$ whichever is lower	−15 dBm	1 MHz
$(F_{c2} + 16\,\text{MHz})$ or $(F_u + 3.2\,\text{MHz})$ whichever is lower − $(F_{c2} + 19.2\,\text{MHz})$ or $(F_u + 3.2\,\text{MHz})$ whichever is lower	−25 dBm	1 MHz
$(F_{c2} + 19.2\,\text{MHz})$ or $(F_u + 3.2\,\text{MHz})$ whichever is lower − 12.75 GHz	−30 dBm	1 MHz

Table 7.24. Spurious emission limits for TDD BSs for the coexistence with UTRA FDD BSs

Band	Maximum level	Measurement bandwidth	Note
$1920 - 1980\,\text{MHz}$	$-32\,\text{dBm}$	$1\,\text{MHz}$	if UTRA FDD BSs are located in the same geographic area
$2110 - 2170\,\text{MHz}$	$-52\,\text{dBm}$	$1\,\text{MHz}$	if UTRA FDD BSs are located in the same geographic area
$1920 - 1980\,\text{MHz}$	$-86\,\text{dBm}$	$1\,\text{MHz}$	for colocated UTRA FDD BSs
$2110 - 2170\,\text{MHz}$	$-52\,\text{dBm}$	$1\,\text{MHz}$	for colocated UTRA FDD BSs

Table 7.25. Abbreviations used for the description of the UTRA TDD RF-characteristics

DPCH_E_C	Average energy per chip of a DPCH
$\sum \text{DPCH\_E}_C / I_{or}$	The ratio of the sum of all DPCH E_Cs for one service in case of multi-code transmission to the total transmit power spectral density of the DL at the BS antenna connector
$\hat{I}_{or}$	Received (DL) power spectral density measured at the MS antenna connector
I_{or}	Total DL transmit power spectral density at the base station antenna connector
I_{oac}	Power spectral density of the adjacent channel measured at the MS antenna connector
I_{ouw}	Power level (or power spectral density in case of a modulated signal) of the unwanted signal measured at the MS antenna connector

[129]. From the logical channel (see section 6.2.1) DTCH (a logical channel that is mapped in the MAC layer to a traffic channel), a block consisting

Table 7.26. Parameters for the 12.2 kbps DL reference measurement channel for the 3.84 Mcps TDD option

Information bit rate	12.2 kbps
DPCH physical bit rate	47.2 kbps
Midamble	512 chips
Spreading factor	16
TFCI	16 Bit/user/10 ms
Coding	Convolutional coding
Code rate	1/3

Table 7.27. Parameters for the 12.2 kbps DL reference measurement channel for the 1.28 Mcps TDD option

Information bit rate	12.2 kbps
DPCH physical bit rate	32.8 kbps
Midamble	144 chips
Spreading factor	16
TFCI	16 Bit/user/10 ms
Power control	4 Bit/user/10 ms
Coding	Convolutional coding
Code rate	1/3

of 244 information bits is extended with 16 CRC bits and 8 tail bits every 20 ms. After convolutional coding with rate 1/3, this block of 268 bits results in 804 bits which is segmented into two radio frames with 402 bits each. After rate matching by means of puncturing, we receive 382 bits per radio frame to which 90 bits are added from the logical channel DCCH (a point-to-point bi-directional logical channel to carry dedicated control information from higher layers between a MS and the UTRAN). This yields 472 bits per radio frame and thus, 47.2 kbps. These 472 bits are mapped onto *two bursts (type 1) to be transmitted in parallel.* Both bursts feature a midamble of length 512 chips. One burst carries two times 114 information bits and two times 8 TFCI bits. The other burst consists of two times 122 bits. Because of the QPSK modulation format and the definition of the SF as the ratio of chips per symbol, the 122 bits before and after the midamble in both bursts correspond to 976 chips for a SF of 16. Note that *only one slot per radio frame* is allocated for data transmission of the 12.2 kbps reference measurement channel. This is in contrast to the FDD mode where the whole radio frame is occupied by the reference measurement channel, which results in a much lower SF for the TDD mode to carry the same data rate. The 90 bits per radio frame to be physically transmitted from the DCCH are constructed from 100 DCCH information bits (including 4 bits of MAC header) per 40 ms. After adding 12 CRC bits and 8 tail bits, a convolutional coder with rate 1/3 is applied, yielding 360 bits. These are segmented into four times 90 bits for each radio frame.

In the case of the 1.28 *Mcps option* the steps are identical up to the segmentation into 402 bits per radio frame. Afterwards, puncturing yields 268 bits to which 60 bits from the logical channel DCCH are added. This yields 328 bits per radio frame and thus, 32.8 kbps. Furthermore, 16 TFCI bits, 4 TPC bits, and 4 SS bits are added and the resulting 352 bits are divided into 176 bits for each sub-frame. Two bursts are used for the transmission. One burst consists of 40 information bits, 4 TFCI bits, the 144 chip midamble, 2 SS bits, 2 TPC bits, 4 TFCI bits, and 36 information bits (see Fig. 7.9). In the second burst 44 information bits are transmitted before and after the

Table 7.28. TDD reference sensitivity test case

$\sum \text{DPCH\_E}_C/I_{or}$	0 dB
$\hat{I}_{or}$	$-105\,\text{dBm}/3.84\,\text{MHz}$ (3.84 Mcps option)
	$-108\,\text{dBm}/1.28\,\text{MHz}$ (1.28 Mcps option)

midamble, which equals 352 chips owing to the SF of 16 and the QPSK modulation.

Reference Sensitivity Test Case. The reference sensitivity test case specifications are given in Table 7.28. Because of the TDD mode, the transmitter does not operate in this (or any other) receiver test case. For this and all other receiver test cases (except for the maximum input level test case) it is assumed that no other channels are received by the MS besides the wanted DPCH(s).

If we repeat the same simple calculation of the required noise figure for the whole receiver chain, as was performed for the FDD mode, we find for the 3.84 Mcps option under the same assumptions (a necessarily effective $(E_b/N_t)_{\text{eff}}$ of 7.2 dB if the baseband implementation loss is already included and a coding gain G_c of 4 dB):

$$\begin{aligned}
\text{NF} &= P_{\text{DPCH}} + G_p + G_c - \left(\frac{E_b}{N_t}\right)_{\text{eff}} - P_N \\
&= -105\,\text{dBm} - 3\,\text{dB} + 12\,\text{dB} + 4\,\text{dB} - 7.2\,\text{dB} + 108\,\text{dBm} \\
&= 8.8\,\text{dB}.
\end{aligned} \tag{7.9}$$

The subtraction of 3 dB from P_{DPCH} results from the parallel transmission of two bursts in the 12.2 kbps DL reference measurement channel, because we assume the same transmission power for each burst. The resulting NF of 8.8 dB is identical to the NF in the FDD mode. A NF of 10.6 dB results for the 1.28 Mcps option owing to the noise power of $-112.8\,\text{dBm}$ caused by the reduced bandwidth.

If we repeat the computations and estimations performed in the FDD mode for the following TDD receiver test cases, we find the same results (for the 3.84 Mcps mode) as in the FDD mode.

Maximum Input Level Test Case. The maximum input level test case specifies the maximum received input power at the MS antenna connector which should not degrade the specified BER performance of 10^{-3}. The parameters are listed in Table 7.29. For this test case, the difference between the total received power spectral density in the receive band and that of the sum of DPCHs is 7 dB and, therefore, much larger than for the other test cases. This is to account for the in-band interference coming from other users.

Table 7.29. Maximum input level test case for the TDD mode

$\sum \mathrm{DPCH\_E_C}/I_{\mathrm{or}}$	$-7\,\mathrm{dB}$
$\hat{I}_{\mathrm{or}}$	$-25\,\mathrm{dBm}/3.84\,\mathrm{MHz}$ (3.84 Mcps option) $-25\,\mathrm{dBm}/1.28\,\mathrm{MHz}$ (1.28 Mcps option)

Table 7.30. Parameters for the adjacent channel selectivity test case for the TDD mode

ACS	$33\,\mathrm{dB}$
$\sum \mathrm{DPCH\_E_C}/I_{\mathrm{or}}$	$0\,\mathrm{dB}$
$\hat{I}_{\mathrm{or}}$	$-91\,\mathrm{dBm}/3.84\,\mathrm{MHz}$ (3.84 Mcps option) $-91\,\mathrm{dBm}/1.28\,\mathrm{MHz}$ (1.28 Mcps option)
I_{oac} (modulated)	$-52\,\mathrm{dBm}/3.84\,\mathrm{MHz}$ (3.84 Mcps option) $-54\,\mathrm{dBm}/1.28\,\mathrm{MHz}$ (1.28 Mcps option)

Table 7.31. Parameters for the in-band blocking test case for the 3.84 Mcps TDD option

$\sum \mathrm{DPCH\_E_C}/I_{\mathrm{or}}$	$0\,\mathrm{dB}$
$\hat{I}_{\mathrm{or}}$	$-102\,\mathrm{dBm}/3.84\,\mathrm{MHz}$
I_{blocking} (modulated)	$-56\,\mathrm{dBm}/3.84\,\mathrm{MHz}$ at $\pm10\,\mathrm{MHz}$ away
I_{blocking} (modulated)	$-44\,\mathrm{dBm}/3.84\,\mathrm{MHz}$ at $\pm15\,\mathrm{MHz}$ away

Table 7.32. Parameters for the in-band blocking test case for the 1.28 Mcps TDD option

$\sum \mathrm{DPCH\_E_C}/I_{\mathrm{or}}$	$0\,\mathrm{dB}$
$\hat{I}_{\mathrm{or}}$	$-105\,\mathrm{dBm}/1.28\,\mathrm{MHz}$
I_{blocking} (modulated)	$-61\,\mathrm{dBm}/1.28\,\mathrm{MHz}$ at $\pm3.2\,\mathrm{MHz}$ away
I_{blocking} (modulated)	$-49\,\mathrm{dBm}/1.28\,\mathrm{MHz}$ at $\pm4.8\,\mathrm{MHz}$ away

Adjacent Channel Selectivity Test Case. The parameters for the ACS test case specified for the UTRA TDD mode are listed in Table 7.30. ACS is defined as the attenuation of the adjacent channel power (at $\pm5\,\mathrm{MHz}$ offset for the 3.84 Mcps option and at $\pm1.6\,\mathrm{MHz}$ offset for the 1.28 Mcps option) relative to the wanted channel power and must be better than the defined $33\,\mathrm{dB}$ for power classes 2 and 3.

In-band Blocking. The in-band blocking test case is similar to the adjacent channel selectivity test case but here the unwanted signals with much higher power than the wanted receive signal are 10 and 15 MHz (3.84 Mcps) or 3.2 and 4.8 MHz (1.28 Mcps) away. The parameters for this test case are given in Table 7.31 for the 3.84 Mcps option and in Table 7.32 for the 1.28 Mcps option.

Table 7.33. Parameters for the out-of-band blocking test case for the 3.84 Mcps TDD option

$\sum \text{DPCH\_E}_{\text{C}}/I_{\text{or}}$		0 dB
$\hat{I}_{\text{or}}$		$-102\,\text{dBm}/3.84\,\text{MHz}$
I_{blocking} (CW)	$-44\,\text{dBm}$	$1840\,\text{MHz} < f < 1885\,\text{MHz}$ $1935\,\text{MHz} < f < 1995\,\text{MHz}$ $2040\,\text{MHz} < f < 2085\,\text{MHz}$
$I_{\text{blocking}}^{\,1}$ (CW)	$-44\,\text{dBm}$	$1790\,\text{MHz} < f < 1835\,\text{MHz}^1$ $2005\,\text{MHz} < f < 2050\,\text{MHz}^1$
$I_{\text{blocking}}^{\,2}$ (CW)	$-44\,\text{dBm}$	$1850\,\text{MHz} < f < 1895\,\text{MHz}^2$ $1945\,\text{MHz} < f < 1990\,\text{MHz}^2$
I_{blocking} (CW)	$-30\,\text{dBm}$	$1815\,\text{MHz} < f < 1840\,\text{MHz}$ $2085\,\text{MHz} < f < 2110\,\text{MHz}$
$I_{\text{blocking}}^{\,1}$ (CW)	$-30\,\text{dBm}$	$1765\,\text{MHz} < f < 1790\,\text{MHz}^1$ $2050\,\text{MHz} < f < 2075\,\text{MHz}^1$
$I_{\text{blocking}}^{\,2}$ (CW)	$-30\,\text{dBm}$	$1825\,\text{MHz} < f < 1850\,\text{MHz}^2$ $1990\,\text{MHz} < f < 2015\,\text{MHz}^2$
I_{blocking} (CW)	$-15\,\text{dBm}$	$1\,\text{MHz} < f < 1815\,\text{MHz}$ $2110\,\text{MHz} < f < 12750\,\text{MHz}$
$I_{\text{blocking}}^{\,1}$ (CW)	$-15\,\text{dBm}$	$1\,\text{MHz} < f < 1765\,\text{MHz}^1$ $2075\,\text{MHz} < f < 12750\,\text{MHz}^1$
$I_{\text{blocking}}^{\,2}$ (CW)	$-15\,\text{dBm}$	$1\,\text{MHz} < f < 1825\,\text{MHz}^2$ $2015\,\text{MHz} < f < 12750\,\text{MHz}^2$

---

[1] If operated in the frequency bands $1850 - 1910$ MHz and $1930 - 1990$ MHz in ITU Region 2 which comprises the countries of North and South America.

[2] If operated in the frequency band $1910 - 1930$ MHz in ITU Region 2.

Out-of-band Blocking. In the out-of-band blocking test case the BER must not exceed 10^{-3} in the presence of CW interferers outside the receive band. The specifications are listed in Table 7.33 for the 3.84 Mcps option and a graphical representation is given in Fig. 7.23 for operation in countries outside ITU region 2. For frequencies in the range $1885\,\text{MHz} < f < 1900\,\text{MHz}$, $1920\,\text{MHz} < f < 1935\,\text{MHz}$, $1995\,\text{MHz} < f < 2010\,\text{MHz}$, and $2025\,\text{MHz} < f < 2040\,\text{MHz}$, the appropriate blocking or adjacent channel selectivity test case specification must be fulfilled. For operation in ITU region 2 this is valid for the frequency ranges $1835\,\text{MHz} < f < 1850\,\text{MHz}$ and $1990\,\text{MHz} < f < 2005\,\text{MHz}$ if operated in the frequency bands $1850 - 1910$ MHz and $1930 - 1990$ MHz, and for $1895\,\text{MHz} < f < 1910\,\text{MHz}$ and $1930\,\text{MHz} < f < 1945\,\text{MHz}$ if operated in the frequency band $1910 - 1930\,\text{MHz}$.

Table 7.34 lists the requirements for the 1.28 Mcps option. For frequencies in the range $1895.2\,\text{MHz} < f < 1900\,\text{MHz}$, $1920\,\text{MHz} < f < 1924.8\,\text{MHz}$, $2005.2\,\text{MHz} < f < 2010\,\text{MHz}$, and $2025\,\text{MHz} < f < 2029.8\,\text{MHz}$, the appropriate blocking or adjacent channel selectivity test case specification must be fulfilled. For operation in ITU region 2 this is valid for the frequency

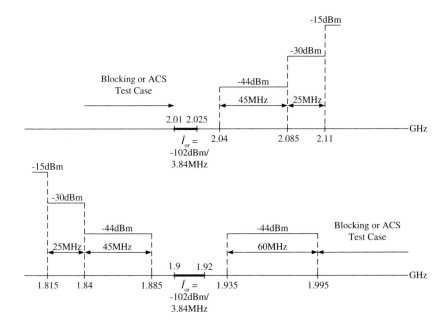

Fig. 7.23. Out-of-band blocking test case for the 3.84 Mcps TDD option for operation outside region 2

ranges 1845.2 MHz $< f <$ 1850 MHz and 1990 MHz $< f <$ 1994.8 MHz if operated in the frequency bands $1850 - 1910$ MHz and $1930 - 1990$ MHz, and for 1905.2 MHz $< f <$ 1910 MHz and 1930 MHz $< f <$ 1945 MHz if operated in the frequency band $1910 - 1930$ MHz.

For this test case, up to 24 exception frequencies are allowed in each assigned frequency channel (when measured using a 1 MHz step size) at which spurious response signals occur and at which the out-of-band blocking specifications need not to be fulfilled. At these exception frequencies the spurious response test case must be met.

Intermodulation Test Case. For the intermodulation test case, two types of interferers are specified: a CW interferer and a (W-CDMA) modulated interference signal. The parameters are given in Tables 7.35 and 7.36.

Spurious Response Test Case. At the exception frequencies of the out-of-band blocking test case (where spurious response signals occur), the specifications listed in Table 7.37 must be met.

Spurious Emissions. The spurious emissions generated by the MS receiver at the antenna connector must stay below the values listed in Table 7.38.

Table 7.34. Parameters for the out-of-band blocking test case for the 1.28 Mcps TDD option

$\sum \mathrm{DPCH\_E_C}/I_{\mathrm{or}}$		0 dB
$\hat{I}_{\mathrm{or}}$		$-105\,\mathrm{dBm}/1.28\,\mathrm{MHz}$
I_{blocking} (CW)	$-44\,\mathrm{dBm}$	$1840\,\mathrm{MHz} < f < 1895.2\,\mathrm{MHz}$ $1924.8\,\mathrm{MHz} < f < 2005.2\,\mathrm{MHz}$ $2029.8\,\mathrm{MHz} < f < 2085\,\mathrm{MHz}$
$I_{\mathrm{blocking}}^{1}$ (CW)	$-44\,\mathrm{dBm}$	$1790\,\mathrm{MHz} < f < 1845.2\,\mathrm{MHz}^{1}$ $1994.8\,\mathrm{MHz} < f < 2050\,\mathrm{MHz}^{1}$
$I_{\mathrm{blocking}}^{2}$ (CW)	$-44\,\mathrm{dBm}$	$1850\,\mathrm{MHz} < f < 1905.2\,\mathrm{MHz}^{2}$ $1934.8\,\mathrm{MHz} < f < 1990\,\mathrm{MHz}^{2}$
I_{blocking} (CW)	$-30\,\mathrm{dBm}$	$1815\,\mathrm{MHz} < f < 1840\,\mathrm{MHz}$ $2085\,\mathrm{MHz} < f < 2110\,\mathrm{MHz}$
$I_{\mathrm{blocking}}^{1}$ (CW)	$-30\,\mathrm{dBm}$	$1765\,\mathrm{MHz} < f < 1790\,\mathrm{MHz}^{1}$ $2050\,\mathrm{MHz} < f < 2075\,\mathrm{MHz}^{1}$
$I_{\mathrm{blocking}}^{1}$(CW)	$-30\,\mathrm{dBm}$	$1825\,\mathrm{MHz} < f < 1850\,\mathrm{MHz}^{2}$ $1990\,\mathrm{MHz} < f < 2015\,\mathrm{MHz}^{2}$
I_{blocking} (CW)	$-15\,\mathrm{dBm}$	$1\,\mathrm{MHz} < f < 1815\,\mathrm{MHz}$ $2110\,\mathrm{MHz} < f < 12750\,\mathrm{MHz}$
$I_{\mathrm{blocking}}^{1}$ (CW)	$-15\,\mathrm{dBm}$	$1\,\mathrm{MHz} < f < 1765\,\mathrm{MHz}^{1}$ $2075\,\mathrm{MHz} < f < 12750\,\mathrm{MHz}^{1}$
$I_{\mathrm{blocking}}^{2}$(CW)	$-15\,\mathrm{dBm}$	$1\,\mathrm{MHz} < f < 1825\,\mathrm{MHz}^{2}$ $2015\,\mathrm{MHz} < f < 12750\,\mathrm{MHz}^{2}$

[1] If operated in the frequency bands $1850 - 1910$ MHz and $1930 - 1990$ MHz in ITU Region 2 which comprises the countries of North and South America.

[2] If operated in the frequency band $1910 - 1930$ MHz in ITU Region 2.

Table 7.35. Parameters for the intermodulation test case for the 3.84 Mcps TDD option

$\sum \mathrm{DPCH\_E_C}/I_{\mathrm{or}}$	0 dB
$\hat{I}_{\mathrm{or}}$	$-102\,\mathrm{dBm}/3.84\,\mathrm{MHz}$
I_{ouw1} (CW)	$-46\,\mathrm{dBm}$ at 10 MHz away
I_{ouw2} (modulated)	$-46\,\mathrm{dBm}/3.84\,\mathrm{MHz}$ at 20 MHz away

7.8.5 BS Receiver Test Cases for the TDD Mode

Unless otherwise stated all parameters are specified at the antenna connector of the BS. They are defined using the 12.2 kbps UL reference measurement channel with the specifications listed in Table 7.39 for the 3.84 Mcps option and in Table 7.40 for the 1.28 Mcps option. If not otherwise stated, the coded bit error rate under the conditions given for each BS receiver test case must not exceed 10^{-3}.

Table 7.36. Parameters for the intermodulation test case for the 1.28 Mcps TDD option

$\sum \text{DPCH\_E}_{\text{C}}/I_{\text{or}}$	0 dB
$\hat{I}_{\text{or}}$	-105 dBm/1.28 MHz
I_{ouw1} (CW)	-46 dBm at 3.2 MHz away
I_{ouw2} (modulated)	-46 dBm/1.28 MHz at 6.4 MHz away

Table 7.37. Parameters for the TDD spurious response test case

Wanted signal level	-102 dBm/3.84 MHz -105 dBm/1.28 MHz
Unwanted signal level	-44 dBm at spurious response frequencies

Table 7.38. Parameters for the TDD spurious emissions test case

Band	Maximum level	Measurement bandwidth
9 kHz – 1 GHz	-57 dBm	100 kHz
1 GHz – 1.9 GHz 1.92 GHz – 2.01 GHz 2.025 GHz – 2.11 GHz	-47 dBm [1]	1 MHz
1.9 GHz – 1.92 GHz 2.01 GHz – 2.025 GHz 2.11 GHz – 2.17 GHz	-60 dBm[2] -64 dBm[3]	3.84 MHz (3.84 Mcps option) 1.28 MHz (1.28 Mcps option)
2.17 GHz – 12.75 GHz	-47 dBm	1 MHz

[1] With the exception of frequencies between 12.5 MHz/4 MHz below the first carrier frequency and 12.5 MHz/4 MHz above the last carrier frequency used by the MS for the 3.84 Mcps/1.28 Mcps option, respectively.

[2] With the exception of frequencies between 12.5 MHz below the first carrier frequency and 12.5 MHz above the last carrier frequency used by the MS.

[3] With the exception of frequencies between 4 MHz below the first carrier frequency and 4 MHz above the last carrier frequency used by the MS.

Reference Sensitivity Test Case. The BS reference sensitivity level is -109 dBm and -110 dBm for the 3.84 Mcps and the 1.28 Mcps option, respectively.

Dynamic Range. The BER of 10^{-3} must be fulfilled for a wanted signal level of -79 dBm (reference sensitivity level $+30$ dB) in the presence of an interferer (AWGN) level of -73 dBm/3.84 MHz for the 3.84 Mcps option. For the 1.28 Mcps option, the values change to -80 dBm for the wanted signal level in the presence of an interferer with -76 dBm/1.28 MHz.

Adjacent Channel Selectivity Test Case. The parameters for this test case are listed in Tables 7.41 and 7.42 for the 3.84 Mcps and 1.28 Mcps options, respectively.

Table 7.39. Physical parameters for the 12.2 kbps UL reference measurement channel for the 3.84 Mcps TDD option

Information bit rate	12.2 kbps
DPCH physical bit rate	45.2 kbps
Spreading factor	8
Midamble	512 chips
TFCI	16 bit/user
Power control	2 bit/user
Coding	Convolutional coding
Code rate	1/3

Table 7.40. Physical parameters for the 12.2 kbps UL reference measurement channel for the 1.28 Mcps TDD option

Information bit rate	12.2 kbps
DPCH physical bit rate	32.8 kbps
Spreading factor	8
Midamble	144 chips
TFCI	16 bit/user/10 ms
Power control	4 bit/user/10 ms
Coding	Convolutional coding
Code rate	1/3

Table 7.41. Parameters for the BSs adjacent channel selectivity test case for the 3.84 Mcps TDD option

Wanted signal	−103 dBm
Interfering signal	−52 dBm/3.84 MHz
Frequency offset	5 MHz

Table 7.42. Parameters for the BSs adjacent channel selectivity test case for the 1.28 Mcps TDD option

Wanted signal	−104 dBm
Interfering signal	−55 dBm/1.28 MHz
Frequency offset	1.6 MHz

Blocking Characteristics. The blocking characteristics for the 3.84 Mcps TDD option is specified in Table 7.43 for the frequency band A (see Table 7.1), in Table 7.44 for the frequency band B, and in Table 7.45 for the frequency band C. The parameters for the 1.28 Mcps TDD option are listed in Table 7.46 for the frequency band A, in Table 7.47 for the frequency band B, and in Table 7.48 for the frequency band C.

Table 7.43. Parameters for the blocking characteristic in TDD mode (3.84 Mcps option) for the frequency band A

Center frequency of interfering signal	Interfering signal level	Wanted signal level	Minimum offset of interferer	Type of interfering signal
1880 − 1980 MHz 1990 − 2045 MHz	−40 dBm	−103 dBm	10 MHz	W-CDMA signal with one code
1 − 1880 MHz 1980 − 1990 MHz 2045 − 12750 MHz	−15 dBm	−103 dBm	–	CW carrier

Table 7.44. Parameters for the blocking characteristic in TDD mode (3.84 Mcps option) for the frequency band B

Center frequency of interfering signal	Interfering signal level	Wanted signal level	Minimum offset of interferer	Type of interfering signal
1830 − 2010 MHz	−40 dBm	−103 dBm	10 MHz	W-CDMA signal with one code
1 − 1830 MHz 2010 − 12750 MHz	−15 dBm	−103 dBm	–	CW carrier

Table 7.45. Parameters for the blocking characteristic in TDD mode (3.84 Mcps option) for the frequency band C

Center frequency of interfering signal	Interfering signal level	Wanted signal level	Minimum offset of interferer	Type of interfering signal
1890 − 1950 MHz	−40 dBm	−103 dBm	10 MHz	W-CDMA signal with one code
1 − 1890 MHz 1950 − 12750 MHz	−15 dBm	−103 dBm	–	CW carrier

If UTRA TDD BSs are co-located with GSM900 and/or DCS1800 BSs, additional blocking requirements are necessary for the protection of the TDD BS receiver. Table 7.49 lists these additional requirements.

Intermodulation Test Case. The parameters for the intermodulation test case are given in Table 7.50. Both interfering signals are fed into the BS antenna input together with the wanted signal. Under these conditions the BER must stay below 10^{-3}.

Spurious Emissions. The spurious emissions generated by the BS receiver at the antenna connector must stay below the values listed in Table 7.51. The requirements apply to all BSs with separate receive and transmit antenna

Table 7.46. Parameters for the blocking characteristic in TDD mode (1.28 Mcps option) for the frequency band A

Center frequency of interfering signal	Interfering signal level	Wanted signal level	Minimum offset of interferer	Type of interfering signal
1880 − 1980 MHz 1990 − 2045 MHz	−40 dBm	−104 dBm	3.2 MHz	W-CDMA signal with one code
1 − 1880 MHz 1980 − 1990 MHz 2045 − 12750 MHz	−15 dBm	−104 dBm	–	CW carrier

Table 7.47. Parameters for the blocking characteristic in TDD mode (1.28 Mcps option) for the frequency band B

Center frequency of interfering signal	Interfering signal level	Wanted signal level	Minimum offset of interferer	Type of interfering signal
1830 − 2010 MHz	−40 dBm	−104 dBm	3.2 MHz	W-CDMA signal with one code
1 − 1830 MHz 2010 − 12750 MHz	−15 dBm	−104 dBm	–	CW carrier

Table 7.48. Parameters for the blocking characteristic in TDD mode (1.28 Mcps option) for the frequency band C

Center frequency of interfering signal	Interfering signal level	Wanted signal level	Minimum offset of interferer	Type of interfering signal
1890 − 1950 MHz	−40 dBm	−104 dBm	3.2 MHz	W-CDMA signal with one code
1 − 1890 MHz 1950 − 12750 MHz	−15 dBm	−104 dBm	–	CW carrier

ports and must be tested if both transmitter and receiver are operating, and with the transmit port terminated.

Table 7.49. Additional blocking requirements for BSs in the TDD mode in case of co-location with GSM900 and/or DCS1800 BSs for the frequency band A

Center frequency of interfering signal	Interfering signal level	Wanted signal level	Minimum offset of interferer	Type of interfering signal
Co-location with GSM900				
$925 - 960$ MHz	$+16$ dBm	-103 dBm	–	CW carrier
Co-location with DCS1800				
$1805 - 1880$ MHz	$+16$ dBm	-103 dBm	–	CW carrier

Table 7.50. Parameters for the BSs intermodulation test case for the TDD mode

Wanted signal	-103 dBm
3.84 Mcps option	
CW interferer	-48 dBm at ± 10 MHz away
W-CDMA signal with one code	-48 dBm at ± 20 MHz away
1.28 Mcps option	
CW interferer	-48 dBm at ± 3.2 MHz away
W-CDMA signal with one code	-48 dBm at ± 6.4 MHz away

Table 7.51. Parameters for the BS TDD spurious emissions test case

Band	Maximum level	Measurement bandwidth
9 kHz $- 1$ GHz	-57 dBm	100 kHz
1 GHz $- 1.9$ GHz 1.98 GHz $- 2.01$ GHz	-47 dBm[1]	1 MHz
1.9 GHz $- 1.98$ GHz 2.01 GHz $- 2.025$ GHz	-78 dBm[2] -83 dBm[3]	3.84 MHz (3.84 Mcps option) 1.28 MHz (1.28 Mcps option)
2.025 GHz $- 12.75$ GHz	-47 dBm	1 MHz

[1] With the exception of frequencies between 12.5 MHz/4 MHz below the first carrier frequency and 12.5 MHz/4 MHz above the last carrier frequency used by the MS for the 3.84 Mcps/1.28 Mcps options, respectively.

[2] With the exception of frequencies between 12.5 MHz below the first carrier frequency and 12.5 MHz above the last carrier frequency used by the MS.

[3] With the exception of frequencies between 4 MHz below the first carrier frequency and 4 MHz above the last carrier frequency used by the MS.

8. UMTS Transceiver Design Issues

In this chapter we will first review the homodyne and heterodyne architectures (and other possibilities) for both transmitter and receiver in general and in particular with respect to UMTS. A survey of published analog frontends for UMTS follows. Possible future directions in the transceiver design conclude this chapter.

8.1 Receiver Architectures

8.1.1 Heterodyne Receiver

Figure 8.1 shows the heterodyne receiver structure. The RF front-end (or

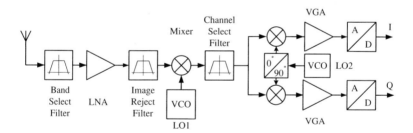

Fig. 8.1. Heterodyne receiver architecture

band select) filter after the antenna suppresses all out-of-band signals (interferers and noise) and performs a first image rejection. Afterwards, an LNA boosts the weak signal before further analog signal processing. The image reject filter in front of the mixer suppresses any signal at the image frequency [43, 134]. In the following, the received signal is translated down to some IF, which is in most cases much lower than the carrier frequency of the RF signal. Channel selection filtering is performed at this IF, which relaxes the requirements for the channel selection filter. Because the wanted signal can be very weak compared to a strong neighboring channel, filters with a high

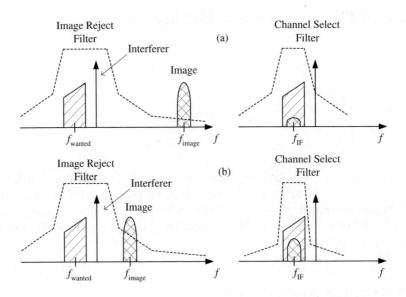

Fig. 8.2. Compromising selectivity vs. sensitivity in a heterodyne receiver: (a) high IF (b) low IF

quality factor (Q) such as surface acoustic wave (SAW) filters or ceramic filters are usually employed for the channel selection [134]. In a final step, the wanted signal centered around the IF is shifted to the baseband by means of a quadrature down-conversion and then amplified, low-pass filtered, and transferred to the digital domain.

The choice of the IF incorporates a principal trade-off between selectivity and sensitivity in the heterodyne receiver design. A comparison of the situations for low and high IF is depicted in Fig. 8.2. With a low IF it is easier to design a highly selective IF filter that attenuates possible interferers near the wanted channel. But the situation is reversed for the image reject filter. It should have a low insertion loss in the passband and a sufficient suppression at the image frequency to yield the highest possible receiver sensitivity. A high IF aids in fulfilling these design goals. To better deal with this selectivity–sensitivity trade-off, more than one IF can be introduced in the receive path which requires the insertion of a further mixer/LO (local oscillator)/channel select filter combination in front of the quadrature down-conversion stage in Fig. 8.1. With this *dual-IF topology* the task of channel selection can be distributed among two filtering stages at different IF frequencies. While this architecture allows great flexibility and enables high performance, it comes at a high cost because of the high number of components. Furthermore, the frequency planning is difficult because of the many possible spurious signals and image bands.

A possible source of distortion in the heterodyne receiver is the *half-IF problem* [45]. When an interferer is received at $(f_{RF} + f_{LO})/2$ (i.e., halfway between the frequency of the desired channel f_{RF} and the LO frequency f_{LO}) it is transferred to $|(f_{RF} + f_{LO}) - 2f_{LO}| = f_{IF}$, if this interferer experiences second-order distortions in the signal path, and if the local oscillator contains a strong second harmonic. Low second-order distortions in the receive path and a clean spectrum of the LO are necessary to alleviate the half-IF problem.

A major advantage of the heterodyne receiver structure is its adaptability to many different receiver requirements. This yields superior performance with respect to selectivity and sensitivity. Therefore, it has been the dominant choice in RF systems for many decades. However, the complexity of the structure and the need for a large number of external components (e.g., the IF filters which in today's cellular phone systems are usually SAW filters [120]) cause problems if a high level of integration is necessary. This is also the major drawback from the cost point-of-view. Furthermore, external components usually have 50 Ohm input and output impedance, as is common for RF components. This requires, e.g., the LNA to drive the low input impedance of the image reject filter, making it difficult to achieve sufficient gain and low noise figure at low power dissipation. Another problem with heterodyne receivers can be the high power consumption required for the amplification at a high IF.

The most important disadvantage of the heterodyne receiver architecture is the missing adaptability of a single design to different wireless standards and modes. Since the external IF filter is optimized for a certain mode of operation, resulting in a fixed bandwidth and center frequency, it cannot be re-used, e.g., for a different mobile communication standard. Mobile terminals for UMTS will most likely be dual-mode devices also supporting a 2G or 2.5G standard. Such a MS would need two different receive paths, each with its own external IF filter, if a heterodyne structure is used. Therefore, designing heterodyne receiver architectures for multi-mode operation would probably result in solutions that are too costly and complex.

8.1.2 Homodyne Receiver

The homodyne receiver architecture (also called zero-IF or direct conversion architecture) avoids the disadvantages of the heterodyne concept by reducing the IF to zero. This saves the (first) mixer, (first) LO, and the channel selection filter at IF, as can be seen from Fig. 8.3 and, moreover, also the image problem vanishes if a quadrature down-converter is used [43]. In fact, the image problem is still present, but its appearance has changed. As the LO frequency is the center frequency of the received channel ($f_{LO} = f_{RF}$), the image is located at $-f_{RF}$. Therefore, the lower half of the spectrum of the wanted channel would fold over the upper half and vice versa if no quadrature down-conversion is applied (except for double-sideband AM signals).

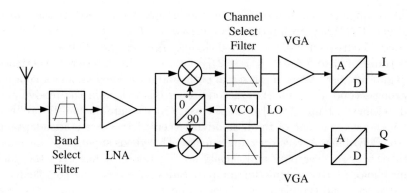

Fig. 8.3. Homodyne receiver architecture

The simplicity of the direct conversion structure offers two important advantages over its heterodyne counterpart. First the problem of image signal suppression is circumvented because $f_{IF} = 0$. As a result, no image filter is required. This may also simplify the LNA (low noise amplifier) design, because there is no need for the LNA to drive a 50 Ohm load which is often necessary when dealing with image rejection filters. Second, the IF SAW filter and subsequent down-conversion stages can be replaced by low-pass filters and baseband amplifiers that can easily be integrated. The possibility of changing the bandwidth of the integrated low-pass filters (and thus adapting the receiver bandwidth) is a major advantage if multi-mode and multi-band applications are of concern.

While known for some time, the homodyne concept was not realized in commercial products despite its obvious advantages until Alcatel adopted this technology in their GSM phones a few years ago [135]. This is due to a number of problems associated with this architecture [43] that do not play a serious role or are at least less serious in a heterodyne receiver.

DC Offset. A serious problem with homodyne receivers is the occurrence of DC-offset voltages that distort the wanted signal after down-conversion (whose spectrum extends to zero frequency) and, more importantly, can saturate the following receiver stages. We find three possible origins for the DC offset, all of them produced by so-called self-mixing. A first source is LO leakage, meaning that the strong LO signal is coupled to the RF port of the mixer or to the input of the LNA. This leakage arises from substrate or capacitive coupling or from coupling via bond wires in the case of an external LO. The leakage signal is mixed with itself, yielding a DC component at the mixer output. If the LO signal leaks to the antenna and is radiated and reflected back from the surroundings, the generated DC offset varies with time if the scattering environment changes. This makes any DC-offset cancellation especially difficult. A second source for a DC offset is strong interferers appearing at the LNA that leak to the LO port of the mixer, leading again to self-mixing. Finally, we can identify second-order distortions in the mixer as the third possible origin of a DC offset. Any strong signal occurring at the mixer input will contribute to the DC voltage if applied to a nonlinearity containing a second order term. If this strong signal contains a slow AM modulation (at frequencies below the bandwidth of the wanted signal), the AM part also appears in the output, thus, generating a *"time-varying DC offset"*. This can easily be seen, if we assume the signal $x(t) = (A + b \cos(2\pi f_1 t)) \cos(2\pi f_0 t)$ at the input of a second-order nonlinearity. A term, $Ab \cos(2\pi f_1 t)$, appears at the output, i.e., the AM component was demodulated at the second-order nonlinearity.

Together, the three described sources for a DC offset easily contribute to some mV of DC voltage. Since a UMTS receiver needs about 80 dB gain, most of which is assigned to the baseband amplifiers, the analog-to-digital converter will be driven into saturation even by such a small DC offset. Therefore, some cancellation technique is required besides a careful design that maximizes mixer port isolation and minimizes second order distortions. Depending on the application, different ways of cancellation are possible. In TDMA-based systems the idle timeslots can be used to detect the offset voltage from self-mixing due to LO leakage and store it in a capacitor for subtracting during the active timeslot. This method is applicable for the UTRA TDD mode only. The simplest possibility to suppress the DC offset in the FDD mode is to introduce a high-pass filter. For narrow-band systems, this method is not feasible because high-pass filters with extremely low cut-off frequencies (0.1% of the data rate) are necessary [136]. This would result in prohibitively large capacitors and in very long time constants that would prevent tracking fast-changing offsets. For wideband systems, however, high-pass filtering is a possible cancellation technique. As the signal band in UTRA extends from DC up to approximately 2 GHz, the loss of some kHz of bandwidth around DC only slightly degrades the system performance. In [137] it was evaluated by means of uncoded BER simulations that a 10 kHz high-pass filter with at least 60 dB of attenuation at 1 kHz (assuming a "time-varying DC offset" up to the maximum Doppler frequency of 1 kHz) produces an E_b/N_0 degradation of only 0.2 to 0.3 dB. More sophisticated DC-offset cancellation techniques are possible, employing digital signal processing [138] or a combination of both analog and digital methods [139].

DC-offset voltages occur in heterodyne receivers as well, but the channel select filter is a band-pass filter that eliminates the DC component produced in the first mixer. Because the quadrature down-converter operates at the much lower IF, any leakage is greatly reduced, and the subsequent baseband amplifier has a much lower gain as in a homodyne receiver. Thus, DC offsets cause no problems in heterodyne receivers.

I/Q Mismatch. Each quadrature down-converter suffers from I/Q mismatch which is a synonym for (a) different gains in both branches of the

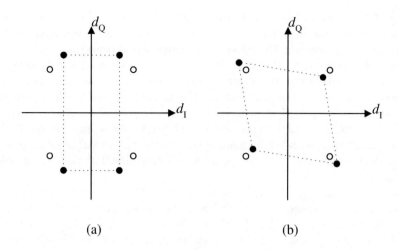

Fig. 8.4. Distortions of the constellation diagram due to I/Q mismatch; (a) gain error (b) phase error

down-conversion and (b) for a deviation from the 90° phase difference between the LO signal in the I and the Q branches. Both effects are always present in any real down-converter circuit. As a consequence, the constellation diagram of the baseband signal is distorted, which increases the BER. A simple calculation [43] shows the effects of gain and phase errors which are displayed in Fig. 8.4. While I/Q mismatch occurs in every quadrature down-converter, for two reasons it is much less of an issue in heterodyne receivers. First, since the down-conversion operates at a much lower frequency, the susceptibility to parasitics is drastically reduced. Second, the signal is already highly amplified before the quadrature down-conversion which makes any mismatches less effective. Besides careful circuit design, a high integration level helps in reducing the I/Q mismatch to values of about 0.25 dB gain mismatch and 1 − 3° phase imbalance with today's technology. If known data bits are transmitted regularly in a system, they can be used to estimate the total I/Q mismatch of the analog circuitry and a correction by means of digital signal processing can be performed. This is an option in W-CDMA-based systems because they use pilot-assisted channel estimation and these pilot symbols can also be used for I/Q mismatch correction.

Even-order Distortion. As already described in section 2.4.1, odd-order nonlinearities are most important in transceiver design because they cause in-band distortions. In direct-conversion receivers, however, even-order nonlinearities (especially of second order) are a critical issue, too. They not only contribute to the DC offset, as already described, but also can generate other in-band distortions in the baseband signal. The mechanism is the following: two closely spaced interferers (at f_1 and f_2), e.g., at the input of the

LNA, which suffers from second-order nonlinearity, generate a signal at the difference frequency $|f_1 - f_2|$ in the output signal of the LNA. Owing to non-negligible feedthrough from the RF to the baseband port of the mixer, part of this signal component appears at the mixer output and corrupts the wanted signal if the difference frequency is below the bandwidth of the wanted signal.

Flicker Noise. If the baseband signal is weak, which is the case in homodyne receivers because of limited amplification by the LNA, $1/f$ noise is a serious problem, especially in CMOS (complementary metal-oxide semiconductor) circuits [136]. In a wideband system like UMTS, the impairments are lower than in narrowband systems. However, for CMOS implementations especially, a careful design of the mixer and the following baseband circuitry is necessary for a successful implementation.

LO Leakage. The LO leakage not only introduces a DC offset but can also be radiated by the antenna. This produces in-band interference for other receivers nearby, e.g., in UTRA FDD this leakage is confined to less than $-60\,\mathrm{dBm}/3.84\,\mathrm{MHz}$ for MSs and less than $-78\,\mathrm{dBm}/3.84\,\mathrm{MHz}$ for BSs.

Channel Selection. After down-conversion to the baseband, channel selection, amplification, and conversion to the digital domain must be performed. These three tasks can be performed in different orders, each of which has other requirements on the amplifier, filter, and ADC (Analog-to-digital converter). A first approach might be to perform the filtering in the digital domain, since it is in general easier to design high performance digital filters than analog ones. But in this case both the amplifier and ADC must feature high linearity and a noise level well below the minimum signal level, which requires high power consumption. If the filtering is performed in the analog domain, we can either amplify before or after the filter. In the first case, a highly linear amplifier with a low noise figure is necessary, which reduces the requisite noise performance of the filter. With the filter before the amplifier, the requirements change and the filter must be highly linear and low in noise. In both cases the dynamic range of the ADC can be greatly reduced compared to the implementation with a digital filter. In a real design, filtering will be divided between the analog and digital domains, and the analog filter will be realized in several stages with amplifiers in between to successively reduce the requirements of the building blocks

The above-listed problems have prevented the widespread use of the direct-conversion receiver architecture up to now. Two factors will probably change this in the near future. First, advances in semiconductor technology today make it possible to design high-performance circuits at low cost, which meets the requirements for homodyne receivers. Second, the need for multi-mode handsets makes the zero-IF structure the premium choice, because it avoids having different RF/IF sections for each standard that the terminal is able to cope with. Instead, programmable analog and digital channel select filters can be designed to deal with the different wireless standards. The multi-mode capability will be essential for any successful commercial product especially for the transition from second to third generation systems. Because of production costs and form-factor requirements, it is very likely that the direct-conversion architecture will be implemented in most wireless transceivers in the near future.

8.1.3 Wideband IF Receiver with Double Conversion

An alternative approach to the homodyne receiver, which is also well suited for monolithic integration, is the wideband IF receiver with double conversion shown in Fig. 8.5 [134, 141]. After the LNA, a quadrature down-conversion of

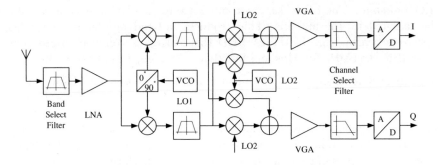

Fig. 8.5. Double conversion wideband IF receiver architecture

all possible receive channels to IF is performed. Therefore, no image problem is present in the case of proper subsequent signal processing. An image reject structure for translating the IF signal to baseband follows, after a wideband low-pass or band-pass filtering (depending on the chosen IF) in both branches. Using this structure, the I and Q components of the wanted signal are added constructively while the image signals are canceled. The local oscillator for the image reject down-conversion (LO2) is tunable in order to properly transfer the wanted channel to the baseband where, finally, the channel selection is performed. The aim of this architecture is to combine the advantages of homo- and heterodyne receivers. The advantages are the following:

- The channel selection is performed in the baseband offering the possibility for multi-standard operation. Since no external high-Q channel select filters are needed, this architecture is well suited for integration.
- The LO leakage signal appears outside the receive band minimizing the problem of a time-varying DC offset.
- The DC offset of the first mixer stage is of no concern since it is out of band and can be removed by a band-pass filter. The DC offset introduced by the image reject mixer is a much smaller problem than in a direct conversion

receiver because of the significantly higher input signal level for the mixer. Furthermore, the DC offset will be fairly constant, allowing the use of some offset cancellation techniques.

- The first LO can be implemented as a fixed frequency oscillator which results in a better phase noise performance.
- Since the second LO operates at a much lower frequency its phase noise performance can be far better than in a homodyne receiver.

The problems of wideband IF receivers are as follows:

- The second LO must be tunable over a fairly wide range of frequencies to be able to translate all possible channels down to baseband. For multi-mode operation, this especially requires a broader than usual tuning range compared to its nominal oscillation frequency.
- Owing to the possibility of strong adjacent channel interference, the image reject down-converter must feature a high dynamic range.
- Any I/Q mismatch limits the image reject capability of the second down-conversion stage which degrades the receiver's sensitivity.

The achievable image rejection (IR) as a function of the phase and gain mismatch is given as [141]

$$
\text{IR}_{\text{dB}} = 10 \log \cdot \left[\frac{1 + (1 + \Delta A)^2 + 2(1 + \Delta A) \cos(\varphi_{\varepsilon 1} + \varphi_{\varepsilon 2})}{1 + (1 + \Delta A)^2 - 2(1 + \Delta A) \cos(\varphi_{\varepsilon 1} - \varphi_{\varepsilon 2})} \right], \quad (8.1)
$$

where $\varphi_{\varepsilon 1}$ and $\varphi_{\varepsilon 2}$ are the deviations from quadrature in the first and second LO, respectively. ΔA is the accumulated gain error between the I and Q branch. To achieve more than 35 dB of image rejection the phase error must stay below $2°$ with perfect gain matching, and the amplitude error must be below 3.6 % at perfect phase matching. Both values are difficult to achieve in mass-production technology. The image rejection capability of the wideband-IF structure can be improved with the help of the RF front-end filter if the IF is chosen high enough (which naturally results when operating the second down-conversion stage at a higher frequency). In this case a total image rejection of 60 to 65 dB can be expected. If we consider the out-of-band blocking test case of the UTRA FDD mode (see section 6.8.4), a CW interferer with -30 dBm must be taken into account when the IF is below 85 MHz – otherwise, the interferer can have a power of -15 dBm. The power spectral density of the wanted signal (DPCH_E_C) is specified to be -114 dBm/3.84 MHz and the total received power spectral density in the receive channel is -103.7 dBm/3.84 MHz. If we make the same assumptions as in the reference sensitivity test case (25 dB of despreading and coding gain), the allowed interference level is -96.2 dBm. The interference power level of -15 dBm is lowered to -36 dBm due to the despreading in the receiver. Therefore, an interference rejection of at least 60 dB is required.

The required dynamic range of the second mixing stage is an even more serious problem of wideband IF receivers than the limited image rejection capability. If we consider the ACS test case, the power of the adjacent channel is specified to be about 50 dB higher than the power of the wanted user signal level (DPCH_E$_C$). The sum of both signals (and all other possible unwanted received signals) must be processed by both mixer stages without distorting the small wanted signal too much. This fact, and the limited image rejection capability, does not make the wideband IF architecture the first choice for an UMTS receiver.

8.1.4 Low-IF Receiver

If the IF in the receiver is low enough, it is possible to convert the IF signal to the digital domain and perform the final down-conversion in the DSP (Digital Signal Processor) [140, 142, 143]. The structure of such a low-IF receiver is shown in Fig. 8.6 and is very similar to the double conversion wideband IF receiver. The main difference is due to the operation of the ADC at IF.

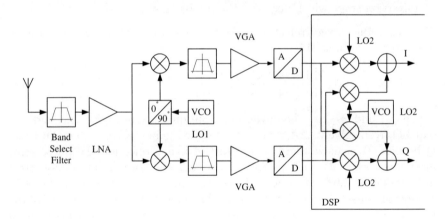

Fig. 8.6. Low-IF receiver architecture

This enables the image reject down-conversion to be implemented digitally and, therefore, without I/Q mismatch. A further difference compared with the wideband IF architecture is the choice of the IF. While the IF in the latter structure is typically high, it is chosen to be as low as possible in low-IF systems to relax the requirements for the ADC. DC offsets are of no concern if band-pass filtering is applied after the first mixer stage and the digital image reject down-conversion does not suffer from any self-mixing and leakage problems.

A disadvantage of low-IF receivers is the required high image suppression [142]. The wanted signal is also the image in a zero-IF receiver, while in a low-IF receiver, the image signal can be substantially stronger than the wanted signal. For a low-IF UMTS receiver, an IF at 2.5 MHz is chosen, which is the lowest possible value, to minimize the ADC requirements. In this case the adjacent channel signal is also the image signal. According to the ACS test case, this signal has about 50 dB higher power than the wanted user signal level (DPCH_E_C). The signal-to-interference ratio (with interference considered as coming only from non-perfect image suppression) in zero-IF receivers is basically equal to the image suppression ratio of the quadrature mixer. In low-IF receivers the achievable signal-to-interference ratio is equal to the image suppression ratio minus the factor for which the image signal can be stronger than the wanted signal. The image suppression is determined by the matching of the first mixer stage but can be enhanced due to adaptive digital signal processing techniques [144]. Owing to the shift of the final image reject down-conversion to the digital domain, the image reject performance of the low-IF receiver can be better than that of the wideband IF structure. However, the high linearity requirements for the image reject down-converter in the wideband IF receiver is shifted to the ADC in a low-IF receiver. It is necessary for the ADC to transfer all necessary information for image suppression into the digital domain. This requires the sampling of a signal composed of the wanted signal and the image signal, which can be as much as 50 dB higher, according to the ACS test case. The resulting necessary high-performance ADC makes this architecture unsuitable for UMTS receivers with today's technology.

In [140] an analog transceiver front-end for the GSM 1800 MHz band has been reported, which makes use of the low-IF architecture. While only the analog part is described and no specifications for the ADC's are given, the use of the low-IF concept is feasible in this case due to the chosen IF of 100 kHz and the fact that the adjacent channel signal is specified to be only 9 dB above the wanted channel.

8.1.5 Digital-IF Receiver

The second down-conversion and subsequent filtering in the heterodyne receiver architecture shown in Fig. 8.1 can be performed in the digital domain leading to the *digital-IF receiver* architecture [43]. The principal issue in this approach is the performance required from the ADC. To limit the requirements on the ADC, a sufficiently low IF has to be chosen, which makes it impossible to employ band-pass filtering to suppress the image frequency. Thus, an image suppression mixer must be used. The image suppression attainable in today's systems is limited to a range of $30 - 55$ dB. Owing to the high demands on the ADC and the image suppression mixer performance, to the author's knowledge, this architecture has only been published once for 3G terminal applications up to now [145] (see the following section). Nevertheless, it is utilized in base stations where many channels must be received and processed simultaneously.

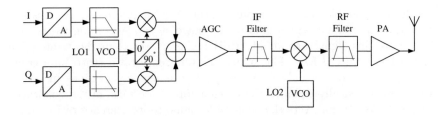

Fig. 8.7. Heterodyne transmitter structure

8.2 Transmitter Architectures

8.2.1 Heterodyne Transmitter

The translation to the carrier frequency is performed in two (or more) steps in a heterodyne transmitter in a similar way to the heterodyne architecture in the receiver. Figure 8.7 shows the schematic of this architecture. Owing to the complex digital modulation, a quadrature up-conversion has to be performed in the first step. An advantage of the heterodyne approach is that since quadrature modulation is performed at low frequencies, I/Q matching is only a minor problem. The band-pass filter after the summation of the I and Q signals suppresses any signal at multiples of the first LO frequency. The final up-conversion is again followed by a band-pass filter to remove the unwanted sideband at the difference frequency $f_{LO2} - f_{LO1}$. The PA boosts the signal to its necessary transmit power. Depending on the duplex technique, either a switch or a duplexer connects the PA to the antenna. Careful design of the switch or duplexer is required, because their insertion loss directly reduces the transmit power and, therefore, the overall efficiency. As in heterodyne receivers, the IF filter is an external device (usually a SAW filter) unsuitable for monolithic integration. A second issue is the choice of the LO frequencies in the case of integrated receive/transmit chips. Here, special care must be taken that LO and IF frequencies, their harmonics, and mixing products, do not appear in the receive band, because any coupling, e.g., via the substrate, would cause interference in the receive path. The spurious emission mask to be fulfilled by the MSs is shown in Fig. 6.32 for the UTRA FDD mode. The design of a frequency plan for an integrated transceiver front-end based on a heterodyne architecture for multi-mode terminals can be a very difficult and even an almost impossible task if the strong clock signals are also taken into account. In [146] a simple calculation of possible spurious signal frequencies for a transceiver designed for UTRA FDD with variable duplex distance and UTRA TDD led to the choice of a homodyne transmitter, in order to avoid unwanted signals in the receive band.

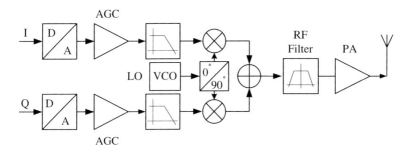

Fig. 8.8. Homodyne transmitter structure

8.2.2 Homodyne Transmitter

In a homodyne transmitter, shown in Fig. 8.8, the up-conversion process is performed in one step by means of a quadrature mixer. While much better suited for integration, there are some difficulties with this architecture. First of all, the local oscillator operates at the same frequency as the high power transmit signal, which has a broad spectrum centered around f_{RF}. This strong transmit signal couples into the oscillator circuit and degrades its phase noise behavior due to so-called *injection pulling*, i.e., a signal injected into an oscillator near its oscillation frequency can change this frequency of operation. If the injected signal is a broadband signal containing AM and PM components as is the case in UMTS, the oscillation frequency is continuously changed. This broadens the oscillator spectrum and degrades the phase noise performance. If the PA is turned on and off periodically, e.g., in UTRA TDD, the problem is even more pronounced. A way of alleviating this issue is the operation of the local oscillator at $2f_{RF}$ with a subsequent division by 2. This approach is commonly used to achieve the necessary LO signals with 90° phase difference for the quadrature conversion with high phase accuracy. A second issue with homodyne transmitters is the I/Q mismatch and the LO leakage of the mixer, both of which produce a continuous wave signal at f_{RF}. This signal appears as an interferer at the center of the transmit spectrum. The I/Q mismatch also directly contributes to the error vector magnitude. Finally, the high required dynamic range of the transmitter is a problem in the direct-conversion architecture, as the output power of a MS transmitter for the UTRA FDD mode ranges from $-50\,\mathrm{dBm}$ to $+24\,\mathrm{dBm}$ and the gain range of the power amplifier is limited to about $30\,\mathrm{dB}$.

8.2.3 Phase-Locked-Loop-Based Transmitters

Instead of performing an I/Q modulation in the transmitter the digital modulation can also be transferred onto the RF carrier by means of magnitude

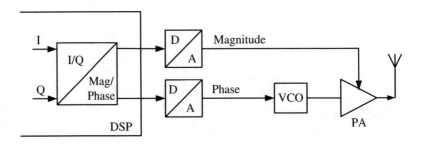

Fig. 8.9. Transmitter structure employing the EER technique

and phase. This concept, also known as *Envelope Elimination and Restoration (EER)* was first proposed by Kahn in 1952 [147]. Figure 8.9 shows the principle. The digital modulation information is separated in an amplitude and a phase part in an DSP. The phase information is used to control the phase/frequency of an oscillator, usually by means of a phase-locked loop (PLL). This constant envelope signal can be amplified in a highly efficient power amplifier. The amplitude information is used to modulate the output amplitude of the phase-modulated RF signal, e.g., by means of changing the PA's power supply. The problems with this method, especially with respect to its applicability to UMTS, are:

- the high amplitude variations (crest factors of 4 and more are possible in the UMTS UL and with several user signals in the DL even higher values can occur) because a linear transfer from the amplitude modulation onto the RF carrier is difficult for low amplitudes and
- the necessary bandwidth of the AM path [148], which usually limits the EER method to narrow-band applications.

Due to these problems no implementations of the EER technique for UMTS applications have been reported in the literature except for [149], but here no specific details are given.

8.3 UMTS Front-End Circuits: Current Developments

In the following we will briefly describe some examples of front-end circuits for UMTS receivers and transmitters.

8.3.1 Homodyne Receivers

The majority of the published work on receiver design is based on the direct conversion topology. The need for high integration has obviously focused concept engineers' attentions on this architecture.

Example R1. An example of a homodyne receiver can be found in [150, 151]. Although not an integrated solution, it is described here for comparison. The whole receiver chain consists of a duplexer, LNA, RF band-pass filter, an even harmonic quadrature mixer for down-conversion [152], and an analog baseband IC [151]. For the duplexer and the LNA, no further data are specified. The down-conversion block is realized using a completely passive even harmonic quadrature mixer composed of two anti-parallel diode pairs as shown in Fig. 8.10. For the proposed mixer configuration, only measure-

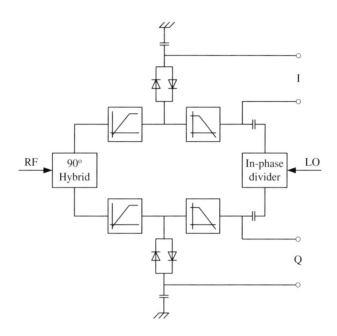

Fig. 8.10. Proposed even harmonic quadrature mixer

ment results from a discrete realization are given. For a LO power of 2 dBm and a load resistance of 220 Ohm at the mixer output, an IIP3 of 0 dBm and a IIP2 of 41 dBm were achieved. The high IIP2 results from the mixer structure, alleviating the problem of the down-conversion of the transmit and LO signals via second order nonlinearities. The baseband IC [151] contains two identical I/Q receive paths to support diversity, each consisting of an LNA with 16 dB gain, a passive RC high-pass to block any DC component, a fifth-order Cauer–Chebyshev low-pass filter, and a cascade of four amplifiers and three attenuators (balanced R-2R ladder networks) with a gain range of 95 dB in steps of 1 dB. It was fabricated with a 0.6 μm BiCMOS process and consumes 20 mA with both receive paths ON and 12 mA with only one path operating. The following performance has been measured with the whole receiver chain (duplexer, LNA, RF band-pass filter, mixer, and analog baseband IC) [150]: a NF of 5.7 dB, an IIP3 of −17.2 dBm, an IIP2 of +26.6 dBm, a sensitivity of −116.4 dBm, an I/Q phase mismatch of 7.7°, and an I/Q amplitude mismatch of 0.4 dB. BER evaluations including a digital demodulator showed the following results: with a chip rate of 4.096 Mchip/s, a symbol rate of 64 ksym/s, and at a channel frequency of 2142 MHz an input power of −116.4 dBm was sufficient to achieve a BER of 10^{-3}. With a setup similar to the intermodulation test case (a CW interferer at 10 MHz and a modulated interferer at 20 MHz offset from the wanted channel, see section 6.8.4) the wanted signal level was −111.8 dBm and the interferer had a maximum power of −45.1 dBm to achieve a BER of 10^{-3}. The allowed adjacent channel power (with −111.8 dBm of wanted signal level) was −46.2 dBm.

Example R2. One of the first prototype chipsets for a W-CDMA direct conversion receiver is described in [153]. The receiver consists of an LNA, mixer, quadrature-phase generation network, baseband amplification and filtering, and ADCs for I and Q path. Four different dies (RF-block, baseband-block, each ADC) have been used to avoid substrate coupling. The ADCs have been realized in 0.5 μm CMOS technology, while a 25 GHz BiCMOS process with a 0.35 μm MOS minimum channel length has been used for the RF and baseband ICs. The receiver is designed to operate in both MS and BS and the channel spacing can be selected digitally between 5 and 20 MHz. The LNA provides a gain of 20 dB with a necessary supply current of 4 mA. The on-chip second order polyphase filter generates the needed quadrature phases of the LO. The measured NF, IIP3, IIP2, and 1dB compression point, all at −5 dBm LO power, are 4 dB (DSB, [double side-band]), −9 dBm, +43 dBm, and −25 dBm, respectively. The I/Q gain and phase imbalance values are below 0.6 dB and below 1°. The analog baseband processing chain includes a pre-amplifier to reduce the noise contribution of the following fifth-order Butterworth active RC filter, output buffers and a servo loop to filter out the DC offset in the input signal. The baseband IC provides a gain range of 78 dB in 3 dB steps. The baseband filter can be programmed to 2, 4, and 8 MHz bandwidth to support chip-rates of 4.096, 8.192, and 16.384 Mcps. To account for process variations, the frequency response of the filter is automatically tuned using 5-bit switched capacitor matrices and a reference time-domain test integrator. The 6-bit 16 Msamples/s ADCs employ a 1.5-bit/stage pipeline architecture with digital correction and interstage gain. The measured NF, IIP3, and IIP2 of the whole receiver are 5.1 dB (DSB), −9.5 dBm, and +38 dBm.

Example R3. The homodyne W-CDMA receiver IC described in [154] consists of, besides an external RF amplifier (NF of 4 dB, 12 dB gain, IIP3 of 11 dBm), a quadrature demodulator with a NF of 13 dB, baseband variable gain amplifiers (VGAs) with a gain control range of 60 dB and an output amplifier, which is also used as active anti-aliasing filter with external R and

C. Between the VGA block and the output amplifier an external LC low-pass filter is inserted in the signal path for channel filtering. For DC-offset cancellation, the mixer output is fed to an offset nulling circuitry via an error integrator. The corner frequency of this high-pass is set at 2 kHz.

Example R4. A promising option is the use of SiGe (silicon germanium) bipolar technology for the receiver front-end. An RFIC (radio frequency integrated circuit) fabricated in SiGe technology described in [155, 156] integrates a complete VCO (voltage controlled oscillator) together with a dual-modulus prescaler, the quadrature phase generation circuitry, the mixer, a low-noise baseband amplifier, and a low-pass blocking filter. Before entering the receiver IC, the signal is fed via a duplexer, an LNA, and an interstage SAW filter that converts the single-ended signal into a differential one. The SAW filter also relaxes the required IIP3 and IIP2 of the following circuits since it further attenuates the transmit signal. The fully integrated VCO operates at twice the RF frequency ($4220 - 4340$ MHz) to reduce LO leakage and enable precision quadrature LO generation and uses octagonal spiral inductors. The low-noise baseband amplifier features two gain settings and is followed by an active Butterworth-type third-order low-pass filter with a corner frequency of 5 MHz. The filter is designed to attenuate blocking signals with frequency offsets larger than 5 MHz, since it provides only little adjacent channel selectivity. The extremely low LO leakage of -95 dBm is remarkably, as is the high IIP2 of 55 dBm of the mixer. This results in a very low DC offset value below 10 mV at the baseband output of the IC. The cascaded values of IIP3 and NF are $+4$ dBm and 15 dBm, respectively, and the I/Q phase mismatch was measured to be $2.5°$. The chip, which draws 33 mA from a 2.7 V supply, is realized in a 0.35 µm SiGe BiCMOS process with an f_T of 75 GHz.

A fully integrated analog baseband IC designed for properly interfacing the above-described zero-IF receiver RFIC is demonstrated in [157]. The partitioning of filtering and amplification was chosen as to simultaneously optimize noise and linearity performance, resulting in cascading the filter blocks with programmable gain amplifiers according to Fig. 8.11. The circuit design of the analog baseband filter is based on extensive system simulation, which especially considered the ACS test case and the blocking specifications, as described in detail in [158]. A thorough investigation led to the design of a seventh-order elliptic filter in combination with a third-order all-pass for phase equalization to meet the in-band distortion limits expressed in terms of error vector magnitude. The elliptic filter was realized by means of a passive pole and three elliptic biquads. The test chip was made using 0.35 µm SiGe BiCMOS technology. The total gain range is 0 dB to 55.5 dB with a resolution of 0.5 dB. The current consumption is 19.5 mA at a bias voltage of 2.7 V.

Example R5. Another example for a direct conversion receiver fabricated with a 0.35 µm 45 GHz f_T SiGe BiCMOS process can be found in [159]. The single-chip receiver includes an LNA, down-conversion mixers, analog channel selection filters, VGAs and 6-bit ADCs. The performance data are an NF of

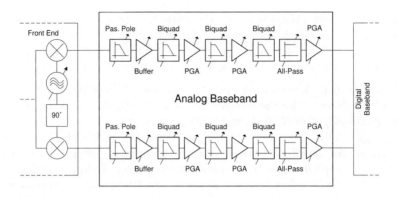

Fig. 8.11. Partitioning of filtering and amplification to simultaneously optimize noise and linearity

3.7 dB, an IIP3 of −16 dBm (high gain), an IIP2 of +18 dBm (high gain), and a 1 dB compression point of −27 dBm. The LNA features a gain of 21 dB and its output signal is AC-coupled to the quadrature mixers to remove the DC component generated by the LNA nonlinearity. The channel selection is performed by a 5th-order Chebyshev type low-pass filter and achieves an adjacent channel attenuation of 36 dB. The VGA has a gain range of 66 dB in 3 dB steps. The 6-bit ADCs use a pipeline architecture and feature a sample frequency of 15.36 Msamples/s. The whole IC consumes only 22 mA from a 2.7 V supply.

8.3.2 Heterodyne Receivers

Example R6. A first UMTS IF transceiver front-end was published in [160] and [161]. Since it is based on the conventional heterodyne architecture it features no multi-mode capability. The chip-set consists of two separate ICs for transmitter and receiver, each with an on-chip synthesizer with integrated VCO tuning and tank. The IF-chips were fabricated in a 0.4 μm/25 GHz silicon bipolar process, operate with 2.7 − 3.3 V supply voltage in an ambient temperature range of −30° to +85°C, and incorporate several power-down modes for power-conscious MS design. The performance of the receiver and the transmitter comply with the ARIB W-CDMA and UMTS standards. The IF receiver IC includes two complete IF paths for antenna diversity and/or service channel monitoring and a common LO generation and distribution. Each path features a three-stage VGA with a gain range of more than 95 dB at the IF frequency of 318 MHz, a quadrature demodulator from the IF down to the baseband, a fifth-order Chebyshev filter, and a first-order all-pass followed by the differential I/Q outputs. The receiver, with both channels and the synthesizer operating, consumes a maximum of 30.2 mA at maximum gain and biased at 2.7 V. Further performance data are a NF of 5 dB, an IIP3 of

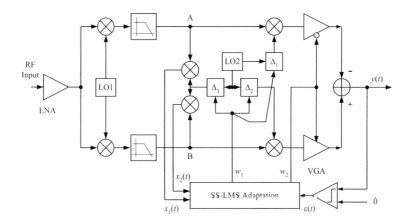

Fig. 8.12. CMOS Weaver-type image-reject receiver with sign–sign LMS calibration

−2 dBm, an I/Q phase mismatch of ±2.5°, an I/Q amplitude mismatch of 0.8 dB, an EVM of 6.2 %, and a SSB phase noise of −132 dBc/Hz at 5 MHz.

8.3.3 Other Receiver Architectures

Example R7. In [162] a *Weaver-type image-reject receiver* fabricated in a 0.25 µm CMOS technology is described. The block diagram is shown in Fig. 8.12. The gain and phase mismatches which are critical for the image rejection capability of this receiver architecture are calibrated using a sign–sign least-mean-square (SS-LMS) algorithm. The mismatches of the second mixer stage are adjusted adaptively and differentially at the IF of 200 MHz, leaving the RF stage undistorted. During the calibration, an image tone is applied at the RF input and the output at the baseband serves as one of the three inputs for the digital SS-LMS circuit. The SS-LMS algorithm adjusts the variable delay stages (denoted by Δ) and the gain adjustment until $y(t)$ approaches zero. A prototype circuit achieved an uncalibrated image-rejection ratio (IRR) of 25 dB which could be improved to 57 dB by means of the SS-LMS calibration. Further data are a NF of 5.2 dB, an IIP3 of −17 dBm, a voltage gain of 41 dB, and a power consumption of 55 mW during calibration and 50 mW in normal receiving mode from a 2.5 V supply.

Example R8. A *single-quadrature receiver* architecture which achieves a maximum IRR of 49 dB without trimming or calibration is demonstrated in [163]. The IC, implemented using 0.2 µm CMOS/SIMOX technology with an f_T of 40 GHz, consists of an LNA, a quadrature mixer, polyphase filters and buffer amplifiers. The polyphase filters are used to convert the external differential LO signal into quadrature signals and to suppress the image signals at the output of the mixers. The single-ended LNA provides a gain of 12.5 dB with a NF of 3.1 dB. The quadrature mixer is designed to suppress the phase errors in the LO signal resulting in lower phase errors in the quadrature IF signal which yields the high *IRR* of > 45 dB for an IF range of 5 to 12 MHz. The IIP3 of the receiver is −15.7 dBm, the NF was measured to be below 10 dB, and the circuit draws 12 mA from a 1 V supply.

Example R9. An integrated CMOS RF front-end for a *digital IF receiver* was published in [145]. The chip consists of an LNA, a mixer, a programmable-gain amplifier (PGA), a fractional-N frequency synthesizer, and a VCO. The IF is located at 190 MHz at which an external SAW filter removes the adjacent channels. Further external components are the RF balun, the loop filter for the RF-PLL and the decoupling capacitors. The cascaded IIP3 of LNA and mixer is −2 dBm, the DSB NF and sensitivity are 3.5 dB and −108 dBm, respectively. With a PGA gain range from −40 to 40 dB switchable in 2 dB steps, the receiver features a gain range of 80 dB. The RF chip draws 52 mA from a 3 V supply.

8.3.4 Comparison of the Described Receivers

Table 8.1 summarizes the performance of the presented receivers. Almost all of the described realizations feature an IIP3 of around −16 dBm which is sufficient to pass the intermodulation test case. Most receivers have a NF between 3.5 dB and about 6 dB. Only for the image-rejection receiver in CMOS/SIMOX technology [163] values of < 10 dB are reported. The SiGe zero-IF receiver from [159], despite including analog channel selection, features a very low NF of only 3.7 dB. However, since not all examples include the analog baseband processing, a fair comparison is difficult. This is also the case if the power dissipation is considered. With 156 mW, the power consumption of the digital IF receiver [145], even without the ADC, is much higher than for all others (except for [153], which is designed for use in both MS and BS and features a much higher IIP3 (−9.5 dBm) than all other implementations), which is a severe drawback for its applicability in mobile terminals. The direct conversion architecture is most favorable concerning the integration. However, only the two-chip solution from [156] and [157] includes VCO and synthesizer. The Weaver-type architecture described in [162] supports the same integration level as the direct conversion receiver. The digital IF receiver described in [145] also includes VCO and synthesizer but requires an IF SAW filter.

Example	R1	R2	R3	R4	R5	R5	R7	R8	R9
IIP3 [dBm]	−17.2	−9.5			−16		−17	−15.7	−16
Mixer IIP3 [dBm]			+12	+13	+18			−3.2	+18
IIP2 [dBm]	+26.6	+38		+55 (mixer)					
NF [dBm]	5.7	5.1	13 (w/o LNA)	15 (w/o LNA)	3.7	5	5.2	< 10	3.5
U [V]		2.7		2.7	2.7	2.7		1	3
I [mA]		128		33	22	30		12	52
P [mW]		346		89	60	81	50	12	156
Architecture			zero-IF			heterod.	Weaver	single-quadrature	digital IF
Building blocks (integrated)	LNA, HPF, LPF, VGA	LNA, Mixer, I/Q gen-eration	Amp., I/Q demod. BB amp.	VCO, Pre-scaler, I/Q-generation mixer, BB amp., blocking filter	LNA, I/Q demod. analog BB filter, VGA ADCs, LO buffer	VGA, I/Q demod. analog BB filter	LNA, I/Q demod. (2×), LPF, amp.	LNA, I/Q demod., polyphase filters, buffer	LNA, mixer, VGA, VCO, synthe-sizer (frac.-N PLL)
Technology	discrete	BiCMOS	SiGe BiCMOS			Si bipolar	CMOS	CMOS/SIMOX	CMOS

Table 8.1. Comparison of the described receivers

8.3.5 Multi-Mode Receivers

Example MMR1. A *dual-band/tri-mode receiver* supporting AMPS, operating in the cellular band between 800 and 900 MHz), IS-95 (CDMA, operating in either the US cellular or the PCS band from 1930 to 1990 MHz) [164], and W-CDMA (UMTS and cdma2000) has been reported in [165]. The RFIC is based on the heterodyne architecture and comprises two independent receiver chains, each consisting of two LNA stages and a mixer with an LO buffer and an IF buffer. The first LNA stage can be bypassed and an external RF filter can be connected between the two LNA stages. One receive path is designed for AMPS and the other can be configured for either IS-95 or W-CDMA. If an external SAW filter with an insertion loss of about 3 dB is used, the minimum NF is 1.8 dB for IS-95 and 2.1 dB for AMPS, the IIP3 is −9 dBm (IS-95) and −6 dBm (AMPS). In the W-CDMA mode the LNA shows a NF of 1.1 dB and an IIP3 of 2 dBm. The RFIC was made using a GaAs (Gallium Arsenide) PHEMT (Pseudomorphic High Electron Mobility Transistor) technology with a minimum gate length of 0.4 μm.

Example MMR2. In [166] a second example for a heterodyne architecture based *dual-band/tri-mode receiver* is described. It supports AMPS, W-CDMA and IS-98-C (CDMA, in either cellular or PCS band). As in example MMR1, two receiver chains are implemented each consisting of an LNA and a mixer. The single VGA after the mixer has inputs for all three modes and is followed by a quadrature demodulator. External filters can be connected before and after the mixer. The LNAs show a NF and an IIP3 of 1.6 dB and 5.6 dBm, respectively, in the cellular band and of 1.8 dB and 3 dBm in the PCS band. The fabrication technology for this RFIC was a 35 GHz Si BiCMOS process.

Example MMR3. The last example in this section features a *direct conversion receiver for dual-band/dual-mode* operation supporting both W-CDMA and GSM [167]. The RFIC contains an LNA, a single-ended to differential converter and a quadrature down-converter. The input transistors and the matching inductors of the LNA are different for both modes. All other on-chip devices are used in both modes. The minimum NF is 4.3 dB (W-CDMA) and 2.3 dB (GSM), the IIP3 is −14.5 dBm (W-CDMA) and −19 dBm (GSM), the IIP2 is 34 dBm (W-CDMA) and 35 dBm (GSM). The chip is fabricated with a 0.35 μm BiCMOS process and consumes 22 mW from a 1.8 V supply.

8.3.6 Heterodyne Transmitters

Example T1. The IF transmitter chip described in [160] and [161] (see also section 8.3.2, example R6) starts with a fifth-order active Butterworth baseband pre-filter both for the I and the Q components. These baseband filters serve as reconstruction filters, minimize ACLR, perform level shifting, and limit the noise bandwidth. The quadrature modulator and a variable gain amplifier with 70 dB gain range at the fixed IF frequency of 285 MHz complete the signal path. The remaining 25 dB gain range has to be realized by means of an RF VGA. The chip includes a synthesizer including integrated VCO tuning and tank with SSB phase noise of -139 dBc/Hz at 5 MHz. The ACLR and EVM of the whole transmit path is 53 dB and 2.7 %, respectively. With an output power of -12 dBm the IF transmitter chip including the synthesizer draws a total of 41.5 mA at maximum gain and biased at 2.7 V.

8.3.7 Homodyne Transmitters

Example T2. A hybrid solution for a homodyne transmitter is presented in [168]. The analog transmit path consists of a low-pass filter for the I and Q paths, an I/Q modulator, an AGC (automatic gain control) amplifier, bandpass filter, and driver and power amplifier. The carrier suppression is better than 40 dB and the ACLR is better than 40 dB with 15 dBm output power. The transmitter is capable of delivering a maximum output power of 24 dBm at the antenna port and controlling the transmit power down to -40 dBm.

8.4 Future Trends

In the following, we will briefly highlight some of the trends for the design and development of future equipment for 3G radio communications. While not claiming completeness, we consider the following topics as the most important ones in the near and mid-term future concerning the hardware developments.

8.4.1 RF CMOS Technology

CMOS technology, which is dominating digital applications, was for a long time considered to be unsuitable for the design of RFICs. This was due to factors such as limited Q values for integrated inductors, the lack of satisfactory device models at GHz frequencies, a limited set of available active and passive devices, the fact that the technology is optimized for digital design, or a worse g_m/I of CMOS transistors compared to bipolar transistors. Advances in CMOS technology, while mainly driven by the enormous impetus from the digital market, pushed the transit frequency of today's 0.13 μm CMOS transistors beyond 50 GHz [169]. This makes the realization of high-speed ICs and RFICs, which were once considered as exclusive domain of III–V and silicon bipolar technologies, becoming feasible in CMOS technology. Together with improved device modeling [170] and the growing number of interconnect layers, which allow for the realization of improved passive components, RF CMOS has become increasingly popular in the last few years. With an integral design approach, taking into account aspects from system level down to device physics, even applications in the 5 GHz range and beyond have been realized, see, e.g., [169]. Advances with respect to improved devices, circuit topologies, and system level architecture make RF CMOS a strong contender for W-CDMA applications [171]. More RF CMOS circuits for WLAN applications have been published (e.g., [172, 173]) than for 3G systems. This is probably due to the considerably higher linearity and noise requirements of 3G systems compared to WLAN systems, a fact arising from the different range and mobility scenarios for the two system types [174]. The majority of the published RFICs for 3G applications are still made within bipolar or BiCMOS technology (see section 8.3.4). However, 3G CMOS solutions have been developed in industry and academia (e.g., [175]) and there is little doubt that in a few years CMOS will have a substantial or even a dominant market share in RF applications for mobile communications.

8.4.2 Low-Power Design

Developing 3G terminals which fulfill or at least nearly approach the standby and talk times of today's 2G and 2.5G MSs at about the same form factor is probably one of the most important and challenging design goals. This is a prerequisite for the full acceptance of 3G systems by the customers. Two items make this goal extremely difficult to achieve for 3G. First, the PA, which determines to a large extent the talk-time of present terminals, has to cope with a linear modulation format (basically QPSK, see section 6.4.4) in 3G systems. The modulation scheme in GSM is GMSK (see section 2.1.5), which features a constant envelope of the RF carrier, allowing for highly nonlinear and, therefore, highly efficient PAs. In UMTS the RF envelope shows high amplitude variations due to the linear modulation scheme and the possibility of transmitting via multiple data channels [176]. This requires a linear PA operation over a wide amplitude range which is only possible with much lower efficiency than in GSM. But a high PA efficiency is extremely important because the transmitter in the UMTS FDD mode is continuously active if not operated in the compressed mode [75]. Linearization techniques [147] can help to improve the PA efficiency. Also strategies like gain switching are possible, which is also applicable for reducing the current consumption of all other amplifiers in the transceiver. The second item that makes a low-power design for 3G terminals a challenge is the fact that the computational load for the digital baseband signal processing rises dramatically in W-CDMA-based 3G systems, owing to the high chip rate and the spread spectrum technology. Despite the advances in process technology for digital circuits this will have a considerable impact on the power consumption of the digital signal processing building blocks. A power conscious design is a multi-disciplinary problem because transceiver architecture and circuit design determine the instantaneous power consumption, while the average power consumption depends to a large extent on a good power management at the system and protocol levels [177, 178].

8.4.3 Integration of RF and Baseband Functionalities

The trend towards a single-chip transceiver appeared some years ago and has stimulated numerous discussions about its realization and practicability. Nevertheless, the steadily increasing level of integration in any type of IC can clearly be identified as a major trend that will continue at least for the next years. This trend also comprises the integration of the analog front-end with the digital baseband. There are numerous obstacles like, e.g., mutual interference caused by substrate coupling, different supply voltage requirements, heat dissipation problems, pin count, etc., but benefits such as reduced component count and required printed circuit board area, as well as enhanced functionality, can be attained and will fuel considerable research effort to overcome these difficulties. This trend for integration of baseband and RF sections also takes place at the system design level. The result will be the enhancement of the RF performance by means of digital signal processing techniques such as digital predistortion for PA linearization [179, 180], improved DC-offset compensation for zero-IF receivers, or the above described on-chip calibration of an image-reject receiver [162] (section 8.3.3).

8.4.4 Software-Defined Radio

Because even for 3G systems a single world-wide standard has not appeared, multi-mode terminals are and will be necessary for world-wide roaming in the next several years. Even if a 3G terminal does not allow world-wide roaming, it has to support a 2G or 2.5G standard at least during the transition phase to 3G. As different standards operate in different frequency bands and require different RF and baseband signal processing, the idea of software-defined radio comes into place [181]. A terminal that can itself adapt by software to different standards depending on the desired application and/or the actual location would greatly reduce costs associated with otherwise necessary separate pieces of hardware for each standard [182]. The most radical approach to software-defined radio would be to digitize the receive signal directly after the antenna and perform all subsequent signal processing in the digital domain with the possibility to adapt this processing by just changing the software. Owing to practical limitations this will certainly not be feasible for wireless communication systems at 2 GHz and above, at least for many years. Much research work is currently going on in the area of suitable software concepts and associated digital hardware structures to realize the baseband signal processing in a software-defined manner [183]. Despite recent advances in digital transceiver design, incorporating digital down-converters and digital up-converters in combination with suitable ADCs and DACs which shift the digital signal processing edge to IF frequencies [184], an RF front-end for a software-defined radio will always remain for some essential RF signal processing like, e.g., anti-aliasing filtering. This analog front-end is often excluded from detailed research and separate RF hardware is assumed for each standard. An important issue for the design of a software-defined radio is the replacement of all narrowband (narrowband in the sense of just covering the frequency band of one standard) fixed frequency components. The duplexer after the antenna, which separates transmit and receive signals in FDD systems, is such a component that has to be removed. Without the duplexer, the strong transmit signal would couple into the highly sensitive receive path and could block the weak wanted receive signal. In [185, 186] a flexible front-end concept is demonstrated for operation in the frequency band from 800 to 2200 MHz with variable channel bandwidths of up to 5 MHz as is required for UMTS. The circumvention of tight RF band filtering as applied in current single-band designs is fundamental to reach this target. A successful application of wideband RF filtering, covering all frequency bands of interest, combined with a subsequent high IF frequency stage and active interference cancellation is demonstrated. The high IF frequency following the RF stage allows image (respectively sideband) rejection. Direct feedthrough of the TX signal into the RX path is greatly suppressed by this new active cancellation technique.

A. Error Function, Complementary Error Function, and Q-Function

If the received signal in a communication system is disturbed by additive white Gaussian noise, its probability density function can be expressed by

$$p_S(x) = \frac{1}{\sqrt{2\pi\sigma^2}} \exp\left\{-\frac{(x-m_s)^2}{2\sigma^2}\right\}. \tag{A.1}$$

Here, x is the amplitude of the received signal (including the noise), m_s is its mean value and σ^2 is the variance ($=$ power) of the noise. From this expression we find the probability that the signal amplitude takes a value between $-\infty$ and x to be

$$P_s(x) = \frac{1}{\sqrt{2\pi\sigma^2}} \int_{-\infty}^{x} \exp\left\{-\frac{(\zeta-m_s)^2}{2\sigma^2}\right\} d\zeta. \tag{A.2}$$

If we perform the following transformation

$$\frac{(\zeta-m_s)}{\sqrt{2\sigma^2}} = t, \quad dt = \frac{d\zeta}{\sqrt{2\sigma^2}}, \quad \int_{-\infty}^{x} \rightarrow \int_{-\infty}^{\frac{(x-m_s)}{\sqrt{2\sigma^2}}},$$

we find

$$P_s\left(\frac{(x-m_s)}{\sqrt{2\sigma^2}}\right) = \frac{1}{\sqrt{\pi}} \int_{-\infty}^{\frac{(x-m_s)}{\sqrt{2\sigma^2}}} \exp\left(-t^2\right) dt$$

$$= \frac{1}{2} \operatorname{erf}\left(\frac{(x-m_s)}{\sqrt{2\sigma^2}}\right) + \frac{1}{2}. \tag{A.3}$$

The *error function* $\operatorname{erf}(x)$ is defined as

$$\operatorname{erf}(x) = \frac{2}{\sqrt{\pi}} \int_{0}^{x} \exp\left(-t^2\right) dt. \tag{A.4}$$

We find the *complementary error function* $\operatorname{erfc}(x)$ to be

$$\operatorname{erfc}(x) = 1 - \operatorname{erf}(x) = 1 - \frac{2}{\sqrt{\pi}} \int_{0}^{x} \exp\left(-t^2\right) dt \tag{A.5}$$

$$= \frac{2}{\sqrt{\pi}} \int_{x}^{\infty} \exp\left(-t^2\right) dt. \tag{A.6}$$

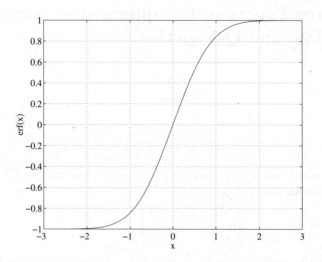

Fig. A.1. Error function

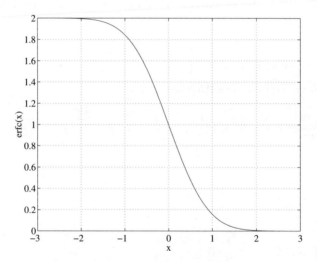

Fig. A.2. Complementary error function

Plots of both functions can be seen in Figs. A.1 and A.2. We note the following relations

$$\operatorname{erf}(x) = -\operatorname{erf}(-x) \tag{A.7}$$

$$1 - \operatorname{erfc}(x) = -1 + \operatorname{erfc}(-x) \tag{A.8}$$

$$1 + \operatorname{erf}(x) = \operatorname{erfc}(-x). \tag{A.9}$$

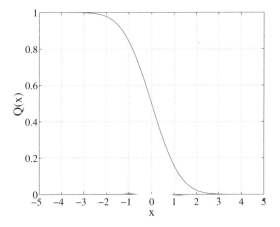

Fig. A.3. Q function

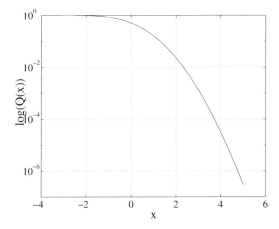

Fig. A.4. Q function in logarithmic scale

The Q-*function* is widely used in digital communications instead of the error function or the complementary error function. It is defined by

$$Q(x) = \frac{1}{2} \operatorname{erfc}\left(\frac{x}{\sqrt{2}}\right) \tag{A.10}$$

$$Q(x) = \frac{1}{\sqrt{2\pi}} \int_x^\infty \exp\left(-\frac{t^2}{2}\right) \mathrm{d}t. \tag{A.11}$$

The Q-function is plotted in Figs. A.3 and A.4

From the above given definitions we can derive further expressions for $P_s\left(\frac{(x-m_s)}{\sqrt{2\sigma^2}}\right)$:

$$P_s\left(\frac{(x-m_s)}{\sqrt{2\sigma^2}}\right) = \frac{1}{\sqrt{\pi}} \int_{-\frac{(x-m_s)}{\sqrt{2\sigma^2}}}^{+\infty} \exp\left(-t^2\right) dt$$

$$= \frac{1}{2}\,\mathrm{erfc}\left(\frac{(m_s-x)}{\sqrt{2\sigma^2}}\right) \tag{A.12}$$

$$P_s(x) = Q\left(\frac{(m_s-x)}{2\sigma}\right). \tag{A.13}$$

In a similar way we compute the probability for the signal amplitude taking values above x. That is

$$1 - P_s(x) = \frac{1}{\sqrt{2\pi\sigma^2}} \int_{-\infty}^{x} \exp\left\{-\frac{(\zeta-m_s)^2}{2\sigma^2}\right\} d\zeta \tag{A.14}$$

$$1 - P_s(x) = Q\left(\frac{(x-m_s)}{2\sigma}\right). \tag{A.15}$$

From this we find that $Q(x)$ is the area under the tail of the Gaussian probability density function (Eq. A.1) with zero mean ($m_s = 0$) and $\sigma^2 = 1$.

B. QoS Parameters for Bearer Services

Table B.1 lists the supported range for BER and maximum transfer delay for different operating environments, valid for both connection and connection-less traffic [101].

Table B.1. Range of QoS parameters to be supported by UMTS

Environment	Real time		Non-real time	
	BER	Max. transfer delay	BER	Max. transfer delay
Satellite [1]	$10^{-3} - 10^{-7}$	$< 400\,\text{ms}$	$10^{-5} - 10^{-8}$	$\geq 1200\,\text{ms}$
Rural outdoor [2]	$10^{-3} - 10^{-7}$	$20 - 300\,\text{ms}$	$10^{-5} - 10^{-8}$	$\geq 150\,\text{ms}$
Urban/suburban outdoor [3]	$10^{-3} - 10^{-7}$	$20 - 300\,\text{ms}$	$10^{-5} - 10^{-8}$	$\geq 150\,\text{ms}$
Indoor/low range outdoor [4]	$10^{-3} - 10^{-7}$	$20 - 300\,\text{ms}$	$10^{-5} - 10^{-8}$	$\geq 150\,\text{ms}$

[1] For terminals with relative speed to ground up to 1000 km/h (plane).
[2] For terminals with relative speed to ground up to 500 km/h.
[3] For terminals with relative speed to ground up to 120 km/h.
[4] For terminals with relative speed to ground up to 10 km/h.

C. UTRA FDD Reference

C.1 Environmental Conditions

Within the given range of the environmental conditions [105] listed in the following, the MS must fulfill the requirements and test cases specified in section 6.8 for the FDD mode and in section 7.8 for the TDD mode [129].

C.1.1 Temperature

Table C.1. Temperature ranges

+15°C to +35°C	For normal conditions (relative humidity of 25% to 75%
−10°C to +55°C	For extreme conditions

C.1.2 Voltage

The MS must fulfill the requirements in the full voltage range (extreme conditions). The lower and higher extreme voltages declared by the manufacturer must not be higher than the lower extreme voltage and lower than the higher extreme voltage specified in Table C.2.

Table C.2. Voltage ranges

Power source	Lower extreme voltage	Higher extreme voltage	Normal conditions voltage
AC mains	0.9 nominal	1.1 nominal	nominal
Regulated lead acid battery	0.9 nominal	1.3 nominal	1.1 nominal
Non-regulated batteries: leclanche/lithium mercury/nickel cadmium	0.85 nominal 0.9 nominal	nominal nominal	nominal nominal

C.1.3 Vibration

The MS must fulfill the requirements if vibrated according to Table C.3.

Table C.3. Vibration

Frequency	Acceleration spectral density (ASD) random vibration
5 Hz to 20 Hz	$0.96\,\mathrm{m^2/s^3}$
20 Hz to 500 Hz	$0.96\,\mathrm{m^2/s^3}$ at 20 Hz, thereafter $-3\,\mathrm{dB/octave}$

C.2 Downlink DPDCH Slot Formats

Table C.4. Downlink DPDCH slot formats

Slot format	Channel bit rate	Channel symbol rate	SF	Bits/ slot	DPDCH bits/slot		DPCCH bits/slot		
	kbit/s	ksym/s			N_{D1}	N_{D2}	N_{TPC}	N_{TFC}	N_{Pilot}
0	15	7.5	512	10	0	4	2	0	4
1	15	7.5	512	10	0	2	2	2	4
2	30	15	256	20	2	14	2	0	2
3	30	15	256	20	2	12	2	2	2
4	30	15	256	20	2	12	2	0	4
5	30	15	256	20	2	10	2	2	4
6	30	15	256	20	2	8	2	0	8
7	30	15	256	20	2	6	2	2	8
8	60	30	128	40	6	28	2	0	4
9	60	30	128	40	6	26	2	2	4
10	60	30	128	40	6	24	2	0	8
11	60	30	128	40	6	22	2	2	8
12	120	60	64	80	12	48	4	8	8
13	240	120	32	160	28	112	4	8	8
14	480	240	16	320	56	232	8	8	16
15	960	480	8	640	120	488	8	8	16
16	1920	960	4	1280	248	1000	8	8	16

C.3 Uplink DPDCH Slot Formats

Table C.5. Uplink DPDCH slot formats

Slot format number	Channel bit rate	Channel symbol rate	SF	Bits/ frame	Bits/ slot	N_{data}
	kbit/s	ksym/s				
0	15	15	256	150	10	10
1	30	30	128	300	20	20
2	60	60	64	600	40	40
3	120	120	32	1200	80	80
4	240	240	16	2400	160	160
5	480	480	8	4800	320	320
6	960	960	4	9600	640	640

C.4 Quantization of the Gain Factors

Signaling values for β_c and β_d	Quantized amplitude ratios β_c and β_d
15	1.0
14	0.9333
13	0.8000
12	0.7333
11	0.6667
10	0.6000
9	0.5333
8	0.4667
7	0.4000
6	0.3333
5	0.2667
4	0.2000
3	0.1333
2	0.0667
1	switched off

Table C.6. Quantization of the gain factors

C.5 Scrambling Code Generation

C.5.1 Uplink Scrambling Code

Long Code. The Gold code used for the construction of the long uplink scrambling sequences is generated from two m-sequences x and y. The x sequence is constructed using the primitive (over GF(2)) polynomial $X^{25} + X^3 + 1$. The y sequence is constructed using the polynomial $X^{25} + X^3 + X^2 + X + 1$. The 24-bit binary representation of the scrambling sequence number n is $n_{23}...n_0$ with n_0 being the least significant bit. The x sequence depends on the chosen scrambling sequence number n and is denoted x_n, in the sequel. Furthermore, let $x_n(i)$ and $y(i)$ denote the ith symbol of the sequence x_n and y, respectively.

The following initial conditions apply:

$x_n(0) = n_0$, $x_n(1) = n_1$,... $x_n(22) = n_{22}$, $x_n(23) = n_{23}$, $x_n(24) = 1$
$y(0) = y(1) = ... = y(23) = y(24) = 1$.

Subsequent symbols are defined recursively:

$x_n(i + 25) = x_n(i + 3) + x_n(i)$ modulo 2, $i = 0, ..., 2^{25} - 27$
$y(i + 25) = y(i + 3) + y(i + 2) + y(i + 1) + y(i)$ modulo 2, $i = 0, ..., 2^{25} - 27$.

The nth long Gold code sequence z_n is then defined as:

$z_n(i) = x_n(i) + y(i)$ modulo 2, $i = 0, 1, 2, ..., 2^{25} - 2$.

Short Code. The short scrambling sequences $c_{\text{short},1,n}(i)$ and $c_{\text{short},2,n}(i)$ are defined from a sequence from the family of periodically extended S(2) codes. The 24-bit binary representation of the code number n is $n_{23}n_{22}...n_0$. The nth quaternary S(2) sequence $z_n(i)$, $0 \leq n \leq 16777215$, is obtained by a modulo 4 addition of a quaternary sequence $a(i)$ and two binary sequences $b(i)$ and $d(i)$. The initial loading of the three sequences is determined by the code number n. The sequence $z_n(i)$ of length 255 is generated as follows:

$z_n(i) = a(i) + 2b(i) + 2d(i)$ modulo 4, $i = 0, 1, ..., 254$.

The quaternary sequence $a(i)$ is generated recursively by the polynomial $g_0(x) = x^8 + x^5 + 3x^3 + x^2 + 2x + 1$:

$a(0) = 2n_0 + 1$ modulo 4
$a(i) = 2n_i$ modulo 4, $i = 1, 2, ..., 7$
$a(i) = 3a(i - 3) + a(i - 5) + 3a(i - 6) + 2a(i - 7) + 3a(i - 8)$ modulo 4,
$i = 8, 9, ..., 254$.

The binary sequence $b(i)$ is generated recursively by the polynomial $g_1(x) = x^8 + x^7 + x^5 + x + 1$:

$b(i) = n_{8+i}$ modulo 2, $i = 0, 1, ..., 7$
$b(i) = b(i - 1) + b(i - 3) + b(i - 7) + b(i - 8)$ modulo 2, $i = 8, 9, ..., 254$.

The binary sequence $d(i)$ is generated recursively by the polynomial $g_2(x) = x^8 + x^7 + x^5 + x^4 + 1$:

$d(i) = n_{16+i}$ modulo 2, $i = 0, 1, ..., 7$
$d(i) = d(i - 1) + d(i - 3) + d(i - 4) + d(i - 8)$ modulo 2, $i = 8, 9, ..., 254$.

The sequence $z_n(i)$ is extended to a length of 256 chips by setting $z_n(255) = z_n(0)$.

C.5.2 Downlink Scrambling Code Generator Polynomials

The Gold code used for the construction of the downlink scrambling sequences is generated from two m-sequences x and y. The x sequence is constructed using the primitive (over GF(2)) polynomial $1 + X^7 + X^{18}$. The y sequence is constructed using the polynomial $1 + X^5 + X^7 + X^{10} + X^{18}$. In the sequel, we denote the sequence depending on the chosen scrambling code number n as z_n. $x(i)$, $y(i)$, and $z_n(i)$ are the ith symbols of the sequence x, y, and z_n, respectively. The m-sequences x and y are constructed as follows:

Initial conditions:

$x(0) = 1$, $x(1) = x(2) = ... = x(16) = x(17) = 0$

$y(0) = y(1) = ... = y(16) = y(17) - 1$.

Subsequent symbols are recursively defined:

$x(i + 18) = x(i + 7) + x(i)$ modulo 2, $i = 0, ..., 2^{18} - 20$

$y(i+18) = y(i+10)+y(i+7)+y(i+5)+y(i)$ modulo 2, $i = 0, ..., 2^{18} - 20$.

The nth Gold code sequence z_n, $n = 0, 1, 2, ..., 2^{18} - 2$, is then defined as:

$z_n(i) = x((i + n)$ modulo $(2^{18} - 1)) + y(i)$ modulo 2, $i = 0, ..., 2^{18} - 2$.

C.6 Synchronization Codes

The PSC C_{psc} is generated by repeating the sequence a modulated by a Golay complementary sequence, and creating a complex-valued sequence with identical real and imaginary components. a is defined as: $\{1, 1, 1, 1, 1, 1, -1, -1, 1, -1, 1, -1, 1, -1, -1, 1\}$. The PSC is defined as $C_{\mathrm{psc}} = (1 + j) \cdot \{a, a, a, -a, -a, a, -a, -a, a, a, a, -a, a, -a, a, a\}$.

The SSCs ($C_{\mathrm{ssc},k}$ with $k = 1, ..., 16$) are complex-valued with identical real and imaginary components and are constructed from position-wise multiplication of a Hadamard sequence and a sequence z, defined as $z = \{b, b, b, -b, b, b, -b, -b, b, -b, b, -b, -b, -b, -b, -b\}$ with $b = a$. a is the same sequence as used in the definition of the PSC. The Hadamard sequences are obtained as the rows in a matrix $\mathbf{H}_{256}$ constructed according to 4.13. The rows are numbered from the top starting with row 0 (the all-ones sequence). The nth Hadamard sequence is the nth row of $\mathbf{H}_{256}$ numbered from the top. $h_m(i)$ and $z(i)$ denote the ith symbol of the sequence h_m and z. From this $C_{\mathrm{ssc},k}$ ($k = 1, 2, 3, ..., 16$) is defined as $C_{\mathrm{ssc},k} = (1 + j) \cdot \{h_m(0) \cdot z(0), h_m(1) \cdot z(1), h_m(2) \cdot z(2), ..., h_m(255) \cdot z(255)\}$ with $m = 16(k - 1)$.

C.7 OCNS

Table C.7. DPCH spreading code, timing offset, and relative level setting for the 16 DPCHs of which the OCNS signal consist

Channelization code	Timing offset($\times 256\,T_{\text{chip}}$)	Level setting (dB)
2	86	-1
11	134	-3
17	52	-3
23	45	-5
31	143	-2
38	112	-4
47	59	-8
55	23	-7
62	1	-4
69	88	-6
78	30	-5
85	18	-9
94	30	-10
102	61	-8
113	128	-6
119	143	0

D. UTRA TDD Reference

D.1 Channelization Code Specific Multiplier

Table D.1. Values for the channelization code specific multiplier

k	$w_{Q=1}^{(k)}$	$w_{Q=2}^{(k)}$	$w_{Q=4}^{(k)}$	$w_{Q=8}^{(k)}$	$w_{Q=16}^{(k)}$
1	1	1	$-j$	1	-1
2		$+j$	1	$+j$	$-j$
3			$+j$	$+j$	1
4			-1	-1	1
5				$-j$	$+j$
6				-1	-1
7				$-j$	-1
8				1	1
9					$-j$
10					$+j$
11					1
12					$+j$
13					$-j$
14					$-j$
15					$+j$
16					-1

D.2 Weight Factors

Table D.2. Weight factors γ_j for different spreading factors SF in the TDD UL

SF of DPCH$_i$	γ_j
16	1
8	$\sqrt{2}$
4	2
2	$2\sqrt{2}$
1	4

D.3 Gain Factors

Table D.3. Possible values for the gain factors β_j in the TDD UL

Signaling value for β_j	β_j
15	16/8
14	15/8
13	14/8
12	13/8
11	12/8
10	11/8
9	10/8
8	9/8
7	8/8
6	7/8
5	6/8
4	5/8
3	4/8
2	3/8
1	2/8
0	1/8

D.4 Synchronization Codes

The primary synchronization codes are constructed as in the FDD mode, described in C.6.

The secondary synchronization codes C_k, with $k = 1, ..., 16$, are complex-valued with identical real and imaginary components and are constructed from position-wise multiplication of a Hadamard sequence and a sequence z, which is defined as $z = \{b, b, b, -b, b, b, -b, -b, b, -b, b, -b, -b, -b, -b, -b\}$ with $b = \{1, 1, 1, 1, 1, 1, -1, -1, -1, 1, -1, 1, -1, 1, 1, -1\}$. The Hadamard sequences are obtained as the rows in the matrix $\mathbf{H}_{256}$ constructed according to (4.13). The rows are numbered from the top starting with row 0 (the all-ones sequence). The nth Hadamard sequence h_n is the nth row of $\mathbf{H}_{256}$. $h_n(i)$ and $z(i)$ denote the ith symbol of the sequence h_n and z. From this C_k ($k = 1, 2, 3, ..., 16$) is defined as $C_k = (1 + j) \cdot \{h_m(0) \cdot z(0), h_m(1) \cdot z(1), h_m(2) \cdot z(2), ..., h_m(255) \cdot z(255)\}$ with $m = 16 \cdot k$. Therefore, each 16th Hadamard sequence is used for constructing a secondary synchronization code.

E. List of Acronyms

2G	Second Generation
3G	Third Generation
3GPP	3rd Generation Partnership Project
AAL2	ATM Adaptation Layer 2
ACLR	Adjacent Channel Leakage Power Ratio
ACS	Adjacent Channel Selectivity
ACTS	Advanced Communications Technology and Services
ADC	Analog-to-digital Converter
AGC	Automatic Gain Control
AI	Acquisition Indicator
AICH	Acquisition Indication Channel
AM	Amplitude Modulation
AMPS	Advanced Mobile Phone Service
AP	Access Preamble
AP-AICH	CPCH Access Preamble Acquisition Indicator Channel
ARIB	Association of Radio Industry and Business
ARQ	Automatic Repeat Request
ASC	Access Service Class
ATM	Asynchronous Transfer Mode
AWGN	Additive White Gaussian Noise
BCH	Broadcast Channel
BEP	Bit Error Probability
BER	Bit Error Rate
BLER	Block Error Rate
BOK	Binary Orthogonal Keying
BPSK	Binary Phase Shift Keying
BS	Base Station

CA	Channel Assignment
CAI	Channel Assignment Indicator
CATT	China Academy of Telecommunication Technology
CCTrCH	Coded Composite Transport Channel
CD/CA-ICH	CPCH Collision Detection/Channel Assignment Indicator Channel
CDI	Collision Detection Indicator
CDMA	Code Division Multiple Access
CD-P	Collision Detection Preamble
CMOS	Complementary Metal-Oxide Semiconductor
CN	Core Network
CPCH	Common Packet Channel
CPFSK	Continuous Phase Frequency Shift Keying
CPICH	Common Pilot Channel
CRC	Cyclic Redundancy Check
CSICH	CPCH Status Indicator Channel
CSMA-CD	Carrier Sense Multiple Access with Collision Detection
CW	Continuous Wave
CWTS	China Wireless Telecommunication Standard Group
DCH	Dedicated Channel
DCCH	Dedicated Control Channel
DECT	Digital Enhanced Cordless Telecommunications
DL	Downlink
DLL	Delay-Locked Loop
DPCCH	Dedicated Physical Control Channel
DPCH	Dedicated Physical Channel
DPDCH	Dedicated Physical Data Channel
DPSK	Differential Phase Shift Keying
DSB	Double Side-band
DSCH	Downlink Shared Channel
DSP	Digital Signal Processor
DS-SS	Direct Sequence Spread Spectrum
DTCH	Dedicated Traffic Channel
DTX	Discontinuous Transmission
DwPCH	Downlink Pilot Channel
DwPTS	Downlink Pilot Timeslot

EDGE	Enhanced Data Rates for GSM Evolution
EER	Envelope Elimination and Restoration
EIRP	Equivalent Isotropic Radiated Power
EMC	Electromagnetic Compatibility
ESA	European Space Agency
ETSI	European Telecommunications Standards Institute
EVM	Error Vector Magnitude
FACH	Forward Access Channel
FBI	Feedback Information
FDD	Frequency Division Duplex
FDMA	Frequency Division Multiple Access
FER	Frame Error Rate
F-FH	Fast Frequency Hopping
FH	Frequency Hopping
FH-SS	Frequency Hopping Spread Spectrum
FIR	Finite Impulse Response
FM	Frequency Modulation
FPACH	Fast Physical Access Channel
FPLMTS	Future Public Land Mobile Telephone System
FSK	Frequency Shift Keying
GaAs	Gallium Arsenide
GMSK	Gaussian Minimum Shift Keying
GP	Guard Period
GPRS	General Packet Radio Service
GPS	Global Positioning System
GSM	Global System for Mobile Communications
HAPS	High Altitude Platform Stations
HIC	Hybrid Interference Cancellation
HPSK	Hybrid PSK
IC	Interference Cancellation
IC	Integrated Circuit
IF	Intermediated Frequency

IIP	Input Referred Intercept Point
IMT-2000	International Mobile Telecommunications
IMT-DS	IMT Direct Spread
IMT-FT	IMT Frequency Time
IMT-MC	IMT Multi Carrier
IMT-SC	IMT Single Carrier
IMT-TC	IMT Time Code
IP	Intercept Point
IP	Internet Protocol
IPn	Intercept Point of nth Order
IPDL	Idle Periods in DL
IR	Image Rejection
IRR	Image-rejection Ratio
ISI	Intersymbol Interference
ISDN	Integrated Services Digital Network
ITU	International Telecommunications Union
JD	Joint Detection
LEO	Low Earth Orbiting
LNA	Low-Noise Amplifier
LO	Local Oscillator
LOS	Line-of-sight
MAC	Medium Access Control
MAI	Multiple Access Interference
MMSE	Minimum Mean Square Error
MOS	Metal-Oxide Semiconductor
MS	Mobile Station
MSK	Minimum Shift Keying
MUD	Multiuser Detection
NADC	North American Digital Cellular
NF	Noise Figure
NLOS	Non-line-of-sight
NMT	Nordic Mobile Telephony

OCNS	Orthogonal Channel Noise Simulator
OF	Orthogonality Factor
OFDM	Orthogonal Frequency Division Multiplexing
OIP	Output Referred Intercept Point
OVSF	Orthogonal Variable Spreading Factor
PA	Power Amplifier
PC	Power Control
PCCC	Parallel Concatenated Convolutional Code
P-CCPCH	Primary Common Control Physical Channel
PCDE	Peak Code Domain Error
PCH	Paging Channel
PC-P	Power Control Preamble
P-CPICH	Primary Common Pilot Channel
PCPCH	Physical Common Packet Channel
PCS	Personal Communication Systems
PDC	Personal Digital Cellular
PDSCH	Physical Downlink Shared Channel
PDU	Protocol Data Units
PHEMT	Pseudomorphic High Electron Mobility Transistor
PHS	Personal Handyphone System
PHY	Physical Layer
PI	Paging Indicator
PIC	Parallel Interference Cancellation
PICH	Paging Indicator Channel
PLL	Phase-locked Loop
PNBSCH	Physical Node B Synchronization Channel
PRACH	Physical Random Access Channel
PSC	Primary Synchronization Code
P-SCH	Primary Synchronization Channel
PSD	Power Spectral Density
PSK	Phase Shift Keying
PUSCH	Physical Uplink Shared Channels
QAM	Quadrature Amplitude Modulation
QoS	Quality of Service
QPSK	Quadrature Phase Shift Keying

RACE	Research and Development of Advanced Communications Technologies in Europe
RACH	Random Access Channel
RADAR	Radio Detection and Ranging
RAN	Radio Access Network
RB	Radio Bearer
RF	Radio Frequency
RFIC	Radio Frequency Integrated Circuit
RITT	Research Institute of Telecommunications Transmission
RLC	Radio Link Control
RMS	Root Mean Square
RRC	Radio Resource Control
RSCP	Received Signal Code Power
RSSI	Received Signal Strength Indicator
RTT	Radio Transmission Technology
SAW	Surface Acoustic Wave
SBP	Special Burst Period
S-CCPCH	Secondary Common Control Physical Channel
SCH	Synchronization Channel
S-CPICH	Secondary Common Pilot Channel
SDMA	Space Division Multiple Access
S-FH	Slow Frequency Hopping
SF	Spreading Factor
SIC	Successive Interference Cancellation
SiGe	Silicon Germanium
SIR	Signal-to-interference Ratio
SMG	Special Mobile Group
SMS	Short Message Service
SNR	Signal-to-noise Ratio
SRB	Signaling Radio Bearer
SS	Spread Spectrum
SS	Synchronization Shift
SSC	Secondary Synchronization Code
S-SCH	Secondary Synchronization Channel
STTD	Space Time Block Coding Transmit Diversity

TA	Timing Advance
TDD	Time Division Duplex
TDL	Tau-Dither Loop
TDMA	Time Division Multiple Access
TD-SCDMA	Time Division Synchronous CDMA
TFC	Transport Format Combination
TFCI	Transport Format Combination Indicator
TFCS	TFC set
TGL	Transmission Gap Length
TH-SS	Time Hopping Spread Spectrum
TIA	Telecommunications Industry Association
TPC	Transmit Power Control
TrCH	Transport Channel
TSTD	Time Switched Transmit Diversity
TTA	Telecommunications Technologies Association
TTC	Telecommunication Technology Committee
TTI	Transmission Time Interval
UARFCN	UTRA Absolute Radio Frequency Channel Number
UE	User Equipment
UL	uplink
ULSC	Uplink Synchronization Control
UMTS	Universal Mobile Telecommunications System
UpPCH	Uplink Pilot Channel
UpPTS	Uplink Pilot Timeslot
USCH	UL Shared Channel
UTRA	UMTS Terrestrial Radio Access
UTRAN	UMTS Terrestrial RAN
UWC	Universal Wireless Communications
UWCC	Universal Wireless Communications Consortium
VAD	Voice Activity Detection
VCO	Voltage Controlled Oscillator
VGA	Variable Gain Amplifier

WARC   World Administrative Radio Conference
W-CDMA   Wideband Code Division Multiple Access
WLANs   Wireless Local Area Networks
WLL   Wireless Local Loop
WRC-2000   World Radiocommunication Conference 2000

ZF-BLE   Zero-Forcing Block Linear Equalizer

References

1. W. Mohr, W. Kohnhäuser, "Access Network Evolution Beyond Third Generation Mobile Communication", *IEEE Communications Magazine*, vol. 38, no. 12, December 2000, pp. 122–133
2. E. Dahlman, B. Gudmundson, M. Nilsson, J. Sköld, "UMTS/IMT-2000 Based on Wideband CDMA", *IEEE Communications Magazine*, vol. 36, no. 9, September 1998, pp. 70–80
3. H. Taub, D. L. Schiling, *Principles of Communication Systems*, McGraw-Hill Series in Electrical Engineering, Communications and Signal Processing, New York, 1986
4. A. B. Carlson, *Communication Systems – An Introduction to Signals and Noise in Electrical Communication*, 3rd Edition, McGraw-Hill Series in Electrical Engineering, Communications and Signal Processing, New York, 1986
5. S. Haykin, *Communication Systems*, 3rd Edition, Wiley, New York, 1994
6. J. G. Proakis, *Digital Communications*, 3rd Edition, McGraw-Hill Series in Electrical and Computer Engineering, New York, 1995
7. A. Papoulis, *The Fourier Integral and its Applications*, McGraw-Hill Classic Textbook Reissue Series, New York, 1962
8. A. V. Oppenheim, A. S. Willsky, S. H. Nawab, *Signals and Systems*, 2nd Edition, Prentice Hall, Engelwood Cliffs, 1997
9. I. Glover, P. Grant, *Digital Communications*, Prentice Hall, London, 1998
10. K. D. Kammeyer, *Nachrichtenübertragung*, B. G. Teubner, Stuttgart, 1996
11. J. M. Wozencraft, I. M. Jacobs, *Principles of Communication Engineering*, Wiley, New York, 1967
12. R. N. McDonough, A. D. Whalen, *Detection of Signals in Noise*, 2nd Edition, Academic Press, San Diego, 1995
13. W. B. Davenport, *Probability and Random Processes*, McGraw-Hill Classic Textbook Reissue Series, New York, 1970
14. T. S. Rappaport, *Wireless Communications*, Prentice Hall, Engelwood Cliffs, 1996
15. Y. Okunev, *Phase and Phase-Difference Modulation in Digital Communications*, Artech House, Norwood, 1997
16. J. Dunlop, D. Girma, J. Irvine, *Digital Mobile Communications and the TETRA System*, Wiley, New York, 1999
17. L. Hanzo, W. Webb, T. Keller, *Single- and Multi-carrier Quadrature Amplitude Modulation: Principles and Applictions for Personal Communications, WLANs and Broadcasting*, Wiley, New York, 2000
18. A. Furuskär, S. Mazur, F. Müller, H. Olofsson, "EDGE: Enhanced Data Rates for GSM and TDMA/136 Evolution", *IEEE Personal Communications*, vol. 6, no. 3, June 1999, pp. 56–66
19. E. A. Lee, D. G. Messerschmitt, *Digital Communication*, 2nd Edition, Kluwer Academic Publishers, Norwell, 1994

20. A. Papoulis, *Probability, Random Variables, and Stochastic Processes*, 3rd Edition, McGraw-Hill Series in Electrical and Computer Engineering, New York, 1991

21. J. C. Liberti, T. S. Rappaport, *Smart Antennas for Wireless Communications: IS-95 and Third Generation CDMA Applications*, Prentice Hall, Engelwood Cliffs, 1999

22. M. Mouly, M.-B. Pautet, *The GSM System for Mobile Communication*, 1992

23. A. Mehrotra, *GSM System Engineering*, Artech House, Norwood, 1997

24. W. C. Y. Lee, *Mobile Communications Design Fundamentals*, 2nd Edition, Wiley, New York, 1993

25. J. Doble, *Introduction to Radio Propagation for Fixed and mMobile Communications*, Artech House, Norwood, 1996

26. K. Siwiak, *Radiowave Propagation and Antennas for Personal Communications*, Artech House, Norwood, 1998

27. B. Sklar, "Rayleigh Fading Channels in Mobile Digital Communiction Systems Part I: Characterization", *IEEE Communications Magazine*, vol. 35, no. 7, July 1997, pp. 90–101

28. B. Sklar, "Rayleigh Fading Channels in Mobile Digital Communiction Systems Part II: Mitigation", *IEEE Communications Magazine*, vol. 35, no. 7, July 1997, pp. 102–109

29. Rappaport T. S., McGillem C. D., "UHF Fading in Factories", *IEEE Journal on Selected Areas in Communications*, vol. 7, no. 1, January 1989, pp. 40–48

30. J. B. Anderson, T. S. Rappaport, S. Yoshida, "Propagation Measurements and Models for Wireless Propagtion Channels", *IEEE Communications Magazine*, vol. 33, no. 1, January 1995

31. G. E. Athanasiadou, A. R. Nix, "A Novel 3-D Indoor Ray-Tracing Propagation Model: The Path Generator and Evaluation of Narrow-Band and Wide-Band Predictions", *IEEE Trans. on Vehicular Technology*, vol. 49, no. 4, July 2000, pp. 1152–1168

32. S. Y. Seidel, T. S. Rappaport, "Site-Specific Propagation Prediction for Wireless In-Building Personal Communication System Design", *IEEE Trans. on Vehicular Technology*, vol. 43, no. 4, November 1994 pp. 879–891

33. U. Dersch, E. Zollinger, "Propagation Mechanisms in Microcell Indoor Environments", *IEEE Trans. on Vehicular Technology*, vol. 43, no. 4, November 1994, pp. 1058–1066

34. G. Liang, H. L. Bertoni, "Review of Ray Modeling Techniques for Site Specific Propagation Prediction", in *Wireless Communications: TDMA versus CDMA* by S. G. Glisic and P. A. Leppänen (eds.), Kluwer, Norwell, 1997, pp. 323–343

35. H. Hashemi, "The Indoor Radio Propagation Channel", *Proc. IEEE*, vol. 81, no. 7, July 1993, pp. 943–968

36. G. L. Stüber, *Principles of Mobile Communication*, Kluwer, Norwell, 1994

37. R. Steele, L. Hanzo, *Mobile Radio Communications*, 2nd Edition, Wiley, New York, 1999

38. S. Ohmori, H. Wakana, S. Kawase, *Mobile Satellite Communications*, Artech House, Norwood, 1998

39. P. W. Baier, J. J. Blanz, R. Schmalenberger, "Fundamentals of Smart Antennas for Mobile Radio Applications", in *Wireless Communications: TDMA versus CDMA* by S. G. Glisic and P. A. Leppänen (eds.), Kluwer, Norwell, 1997, pp. 345–376

40. J. D. Gibson (ed.), *The Mobile Communications Handbook*, CRC Press, Inc., 1996

41. J. A. C. Bingham, *ADSL, VDSL, and Multicarrier Modulation*, Wiley, New York, 2000

42. R. van Nee, R. Prasad, *OFDM for Wireless Multimedia Communications*, Artech House, Norwood, 2000
43. B. Razavi, *RF Microelectronics*, Prentice Hall, Engelwood Cliffs, 1999
44. S. A. Maas, *Nonlinear Microwave Circuits*, IEEE Press, New York, 1997
45. R. C. Sagers, "Intercept Point and Undesired Responses", *IEEE Trans. on Vehicular Technology*, vol. 32, no. 1, February 1983, pp. 121–133
46. P. Vizmuller, *RF Design Guide: Systems, Circuits and Equations*, Artech House, Norwood, 1995
47. H. Pretl, L. Maurer, W. Schelmbauer, R. Weigel, B. Adler, J. Fenk, "Linearity Considerations of W-CDMA Frontends for UMTS", *Proc. 2000 IEEE MTT-S Microw. Symp.*, Boston, USA, July 2000, pp. 433–436
48. R. G. Meyer, A. K. Wong, "Blocking and Desensitization in RF Amplifiers", *IEEE J. of Solid-State Circuits*, vol. 30, no. 8, August 1995, pp. 944–9466
49. R. Perez, *Wireless Communications Design Handbook: Aspects of Noise, Interference, and Environmental Concerns*, Vol. 3 Interference in Circuits, Academic Press, San Diego, 1998
50. G. Gonzales, *Microwave Transistor Amplifiers*, 2nd Edition, Prentice Hall, Engelwood Cliffs, 1997
51. M. Friese, "Multitone Signals with Low Crest Factor", *IEEE Trans. on Communications*, vol. 45, no. 10, October 1997, pp. 1338–1344
52. R. C. Dixon, *Spread Spectrum Systems with Commercial Applications*, 3rd Edition, Wiley, New York, 1994
53. C. E. Shannon, "A mathematical theory of communication", *Bell Syst. Tech. J.*, vol. 27, July 1948, pp. 379–423, October 1948, pp. 623–656
54. S. Verdú, S. W. McLaughlin (eds.), *Information Theory: 50 Years of Discovery*, IEEE Press, New York, 2000
55. R. Prasad, *CDMA for Wireless Personal Communications*, Artech House, Norwood, 1996
56. M. K. Simon, J. K. Omura, R. A. Scholtz, B. K. Levitt, *Spread Spectrum Communications Handbook*, Revised Edition, McGraw-Hill, New York, 1994
57. J. S. Lee, L. E. Miller, *CDMA Systems Engineering Handbook*, Artech House, Boston, 1998
58. Insitute of Electrical and Electronics Engineers, *The IEEE 802.11 Standard for Local and Metropolitan Area Networks*, New York, 1997
59. D. P. Morgan, *Surface-Wave Devices for Signal Processing*, 2nd Edition, Elsevier, Amsterdam, The Netherlands, 1991
60. C. E. Cook, M. Bernfeld, *Radar Signals – An Introduction to Theory and Application*, Academic Press, New York, 1967
61. W. Hirt, S. Pasupathy, "Continuous Phase Chirp (CPC) Signals for Binary Data Communication – Part I: Coherent Detection and Part II: Noncoherent Detection", *IEEE Trans. on Communications*, vol. COM-29, June 1981, pp. 836–858
62. A. Springer, M. Huemer, L. Reindl, C.C.W. Ruppel, A. Pohl, F. Seifert, W. Gugler, R. Weigel, "A Robust Ultra Broadband Wireless Communication System Using SAW Chirped Delay Lines", *IEEE Trans. on Microwave Theory and Techniques*, vol. 46, no. 12, December 1998, pp. 491–494
63. W. Gugler, A. Springer, R. Weigel, "A Robust SAW-Based Chirp – $\pi/4$ DQPSK System for Indoor Applications", *Proc. 2000 IEEE Int. Conf. on Communications (ICC)*, June 2000, New Orleans, USA, pp. 773–777
64. M. Kowatsch, J. T. Lafferl, "A Spread-Spectrum Concept Combining Chirp Modulation and Pseudonoise Coding", *IEEE Trans. on Communications*, vol. COM-31, no. 10, October 1983, pp. 1133–1142

65. S. Glisic, B. Vucetic, *Spread Spectrum CDMA Systems for Wireless Communications*, Artech House, Norwood, 1997
66. B. Sklar, *Digital Communications – Fundamentals and Applications*, Prentice Hall, Engelwood Cliffs, 1988,
67. A. Polydoros, C. L. Weber, "A Unified Approach to Serial Search Spread-Spectrum Code Acquisition", *IEEE Trans. on Communications*, vol. COM-32, no. 5, May 1984, pp. 542–560
68. R. A. Scholtz, "The Origins of Spread-Spectrum Communications", *IEEE Trans. on Communications*, vol. COM-30, no. 5, May 1982, pp. 822–854
69. H. D. Lüke, *Korrelationssignale*, Springer, Berlin, Heidelberg, 1992
70. D. V. Sarwate, M. P. Pursley, "Crosscorrelation Properties of Pseudorandom and Related Sequences", *Proc. IEEE*, vol. 68, no. 5, May 1980, pp. 593–619
71. J. L. Walsh, "A Closed Set of Normal Orthogonal Functions", *American J. Mathematics*, vol. 45, 1923, pp. 5–24
72. M. K. Simon, S. M. Hinedi, W. C. Lindsey, *Digital Communication Techniques*, Prentice Hall, Englewood Cliffs, 1995
73. S. Verdú, *Multiuser Detection*, Cambridge University Press, Cambridge, 1998
74. D. Koulakiotis, H. A. Aghvami, "Data Detection Techniques for DS/CDMA Mobile Systems: A Review", *IEEE Personal Communications*, vol. 7, no. 3, June 2000, pp. 24–34
75. H. Holma, A. Toskala (eds.), *WCDMA for UMTS*, Wiley, New York, 2000
76. T. Ojanperä, R. Prasad (eds.), *Wideband CDMA for Third Generation Mobile Communications*, Artech House, Norwood, 1998
77. J. Cai, D. J. Goodman, "General Packet Radio Service in GSM", *IEEE Communications Magazine*, vol. 35, no. 10, October 1997, pp. 122–131
78. P. Taaghol, B. G. Evans, E. Buracchini, R. De Gaudenzi, G. Gallinaro, J. H. Lee, C. G. Kang, "Satellite UMTS/IMT-2000 W-CDMA Air Interfaces", *IEEE Communications Magazine*, vol. 37, no. 9, September 1999, pp. 116–126
79. R. Bekkers, J. Smits, *Mobile Telecommunications: Standards, Regulation, and Applications*, Artech House, Norwood, 1999
80. J. C. Haartsen, "The Bluetooth Radio System", *IEEE Personal Communications*, vol. 7, no. 1, February 2000, pp. 28–36
81. Institute of Electrical and Electronics Engineers, *Draft Supplement to Standard for Telecommuncations and Information Exchange between Systems – LAN/MAN Specific Requirements – Part 11: Wireless LAN Medium Access (MAC) and Physical Layer (PHY) Specifications: High Speed Physical Layer in the 5 GHz Band*, IEEE P802.11a/D7.0, July 1999
82. UMTS Forum Report No. 9, *The UMTS Third Generation Market – Structuring the Service Revenue Opportunities*, September 2000
83. R. Prasad, W. Mohr, W. Konhäuser (eds.), *Third Generation Mobile Communication Systems*, Artech House, Norwood, 2000
84. S. Nanda, K. Balachandran, S. Kumar, "Adaptation Techniques in Wireless Packet Data Services", *IEEE Communications Magazine*, vol. 38, no. 1, January 2000, pp. 54–64
85. UMTS Forum Report No. 10, *Shaping the Mobile Multimedia Future – An Extended Vision form the UMTS Forum*, 2000
86. http://www.itu.int/newsarchive/wrc2000/IMT-2000/Res-COM5-13.html
87. P. Chaudhury, W. Mohr, S. Onoe, "The 3GPP Proposal for IMT-2000", *IEEE Communications Magazine*, vol. 37, no. 12, December 1999, pp. 72–81
88. ITU Web page: http://www.itu.int/imt/2_rad_devt/proposals/index.html
89. J. F. Huber, D. Weiler, H. Brand, "UMTS, the Mobile Multimedia Vision for IMT-2000: A Focus on Standardization", *IEEE Communications Magazine*, vol. 38, no. 9, September 2000, pp. 129–136

90. ITU Press Release, ITU/99-22, "IMT-2000 Radio Interface Specifications Approved in ITU Meeting in Helsinki", November 5, 1999
91. http://www.3gpp.org
92. http://www.3gpp2.org
93. UMTS Forum Report No. 12, *Naming, Addressing and Identification Issues for UMTS*, December 2000
94. 3GPP, Technical Specification Group (TSG) RAN, "MAC Protocol Specification", *TS 25.321 v4.0.0*, March 2001
95. 3GPP, Technical Specification Group (TSG) SA, "3G Security; Security Architecture", *TS 33.102 v3.4.0*, March 2000
96. 3GPP, Technical Specification Group (TSG) RAN, "RLC Protocol Specification", *TS 25.322 v4.0.0*, March 2001
97. 3GPP, Technical Specification Group (TSG) RAN, "RRC Protocol Specification", *TS 25.331 v4.0.0*, March 2001
98. H. S. Pinto, "UMTS Radio Interface Roll Out Aspects", *IEEE Vehicular Technology Society News*, vol. 48, no. 3, August 2001, pp. 4–9
99. 3GPP, Technical Specification Group (TSG) Services and System Aspects, "Services and Systems Aspects; Network Architecture", *TS 23.002 v4.3.0*, June 2001
100. 3GPP, Technical Specification Group (TSG) Services and System Aspects, "Service Aspects; Service Principles", *TS 22.101 v3.11.0*, October 2000
101. 3GPP, Technical Specification Group (TSG) Services and System Aspects, "Service Aspects; Services and Service Capabilities", *TS 22.105 v3.9.0*, June 2000
102. 3GPP, Technical Specification Group (TSG) RAN WG4, "Physical Channels and Mapping of Transport Channels onto Physical Channels (FDD)", *TS 25.211 v4.0.0*, March 2001
103. 3GPP, Technical Specification Group (TSG) RAN WG4, "Physical Layer Procedures (FDD)", *TS 25.214 v4.0.0*, March 2001
104. 3GPP, Technical Specification Group (TSG) RAN WG4, "UTRA (BS) FDD; Radio Transmission and Reception", *TS 25.104 v4.0.0*, March 2001
105. 3GPP, Technical Specification Group (TSG) RAN WG4, "UE Radio Transmission and Reception (FDD)", *TS 25.101 v4.0.0*, March 2001
106. 3GPP, Technical Specification Group (TSG) RAN WG4, "Spreading and Modulation (FDD)", *TS 25.213 v4.0.0*, March 2001
107. F. Adachi, M. Sawahashi, K. Okawa, "Tree-structured Generation of Orthogonal Spreading Codes with Different Lengths for Forward Link of DS-CDMA Mobile", *Electronics Letters*, vol. 33, no. 1, 1997, pp. 27–28
108. "HPSK Spreading for 3G", *Application Note 1335*, Agilent Technologies, 1999
109. 3GPP, Technical Specification Group (TSG) RAN WG4, "Multiplexing and Channel Coding (FDD)", *TS 25.212 v4.0.0*, December 2000
110. B. Melis, G. Romano, "UMTS W-CDMA: Evaluation of Radio Performance by Means of Link Level Simulations", *IEEE Personal Communications*, vol. 7, no. 3, June 2000, pp. 42–49
111. 3GPP, Technical Specification Group (TSG) RAN WG4, "Radio Resource Management Strategies", *TS 25.922 v4.0.0*, March 2001
112. 3GPP, Technical Specification Group (TSG) RAN WG4, "Interlayer Procedures in Connected Mode", *TS 25.303 v4.0.0*, March 2001
113. 3GPP, Technical Specification Group (TSG) RAN WG4, "Physical Layer - Measurements (FDD)", *TS 25.215 v4.0.0*, March 2001
114. 3GPP, Technical Specification Group (TSG) RAN WG4, "Base Station Conformance Testing (FDD)", *TS 25.141 v4.1.0*, June 2001
115. ITU-R, *Recommendation SM.329-8*

116. 3GPP, Technical Specification Group (TSG) Terminal, "Terminal Conformance Specification; Radio Transmission and Reception (FDD)", *TS 34.121 v3.5.0*, June 2001

117. D. Pimingsdorfer, A. Holm, B. Adler, G. Fischerauer, R. Thomas, A. Springer, R. Weigel, "Impact of SAW RF and IF Filter Characteristics on UMTS Transceiver Performance", *Proc. 1999 IEEE International Ultrasonics Symposium*, Lake Tahoe, USA, October 1999, pp. 365–368

118. TSG-RAN Working Group 4, Nokia Mobile Phones, "MS Receiver Sensitivity in UTRA FDD Mode", *Document TSGW4 #1(99)005*, January 1999

119. O. K. Jensen, et al., "RF Receiver Requirements for 3G W-CDMA Mobile Equipment", *Microwave Journal*, vol. 43, no. 2, February 2000, pp. 22–46

120. L. Maurer, W. Schelmbauer, H. Pretl, Z. Boos, R. Weigel, A. Springer, "Impact of IF-SAW Filtering on the Performance of a W-CDMA Receiver", *Proc. 2000 IEEE Int. Ultrasonics Symposium*, Puerto Rico, October 2000, pp. 375–378

121. L. Maurer, W. Schelmbauer, H. Pretl, A. Springer, B. Adler, Z. Boos, R. Weigel, "Influence of Receiver Frontend Nonlinearities on W-CDMA Signals", *Proc. 2000 Asia-Pacific Microwave Conference*, December 2000, Sidney, Australia, pp. 249–252

122. http://www.3gpp.org/News/mobile_news_2000/ran.htm

123. 3GPP, Technical Specification Group (TSG) RAN WG4, "Physical Channels and Mapping of Transport Channels onto Physical Channels (TDD)", *TS 25.211 v4.0.0*, March 2001

124. 3GPP, Technical Specification Group (TSG) RAN WG4, "Spreading and Modulation (TDD)", *TS 25.223 v4.0.0*, March 2001

125. 3GPP, Technical Specification Group (TSG) RAN WG4, "Multiplexing and Channel Coding (TDD)", *TS 25.222 v4.0.0*, March 2001

126. 3GPP, Technical Specification Group (TSG) RAN WG4, "Physical Layer Procedures (TDD)", *TS 25.224 v4.0.0*, March 2001

127. 3GPP, Technical Specification Group (TSG) RAN WG4, "Synchronisation in UTRAN Stage 2", *TS 25.402 v4.0.0*, March 2001

128. Focus on "Wireless Geolocation Systems and Services", *IEEE Communications Magazine*, vol. 36, no. 4, April 1998

129. 3GPP, Technical Specification Group (TSG) RAN WG4, "UE Radio Transmission and Reception (TDD)", *TS 25.102 v4.0.0*, March 2001

130. 3GPP, Technical Specification Group (TSG) RAN WG4, "UTRA (BS) TDD; Radio Transmission and Reception", *TS 25.105 v4.0.0*, March 2001

131. 3GPP, Technical Specification Group (TSG) RAN WG4, "Requirements for Support of Radio Resources Management; (TDD)", *TS 25.123 v4.0.0*, March 2001

132. 3GPP, Technical Specification Group (TSG) RAN WG4, "Base Station Conformance Testing (TDD)", *TS 25.142 v4.0.0*, March 2001

133. 3GPP, Technical Specification Group (TSG) Terminal, "Terminal Conformance Specification; Radio Transmission and Reception (TDD)", *TS 34.122 v4.0.0*, June 2001

134. S. Mirabbasi, K. Martin, "Classical and Modern Receiver Architectures", *IEEE Communications Magazine*, vol. 38, no. 11, November 2000, pp. 132–139

135. J. Sevenhans, B. Verstraeten, S. Taraborelli, "Trands in Silicon Radio Large Scale Integration: Zero IF Receiver! Zero I & Q Transmitter! Zero Discrete Passives!", *IEEE Communications Magazine*, vol. 38, no. 1, January 2000, pp. 142–147

136. B. Razavi, "Design Considerations for Direct-Conversion Receivers", *IEEE Trans. on Circuits and Systems – II*, vol. 44, no. 6, June 1997, pp. 428–435

137. J. H. Mikkelsen, T. E. Kolding, T. Larsen, T. Klingenbrunn, K. I. Pedersen, P. Mogensen, "Feasibility Study of DC Offset Filtering for UTRA-FDD/WCDMA Direct-Conversion Receiver", *Proc. 17th IEEE NORCHIP Conf.*, Oslo, Norway, November 1999, pp. 34–39

138. A. A. Abidi, "Direct-Conversion Radio Transceivers for Digital Communications", *IEEE J. of Solid-State Circuits*, vol. 30, no. 12, December 1995, pp. 13991410

139. H. Tsurumi, M. Soeya, H. Yoshida, T. Yamaji, H. Tanimoto, Y. Suzuki, "System-Level Compensation Apporach to Overcome Signal Saturation, DC Offset, and 2nd-Order Nonlinear Distortion in Linear Direct Conversion Receiver", *IEICE Trans. Electron.*, vol. E82-C, no. 5, May 1999, pp. 708–716

140. M. S. J. Steyaert, J. Janssens, B. De Muer, M. Borremans, N. Itoh, "A 2-V CMOS Cellular Transceiver Front-End", *IEEE J. of Solid-State Circuits*, vol. 35, no. 12, December 2000, pp. 1895–1907

141. J. C. Rudell, J.-J. Ou, T. B. Cho, G. Chien, F. Brianti, J. A. Weldon, P. R. Gray, "A 1.9-GHz Wide-Band IF Double Conversion CMOS Receiver for Cordless Telephone Applications", *IEEE J. of Solid-State Circuits*, vol. 32, no. 12, December 1997, pp. 2071–2088

142. J. Crols, M. S. J. Steyaert, "Low-IF Topologies for High-Performance Analog Front Ends of Fully Integrated Receivers", *IEEE Trans. on Circuits and Systems – II*, vol. 45, no. 3, March 1998, pp. 269–282

143. J. Crols, M. S. J. Steyaert, "A Single-Chip 900 MHz CMOS Receiver Front-End with a High-Performance Low-IF Topology", *IEEE J. of Solid-State Circuits*, vol. 30, no. 12, December 1995, pp. 1483–1492

144. L. Yu, W. M. Snelgrove, "A Novel Adaptive Mismatch Cancellation System for Quadrature IF Radio Receivers", *IEEE Trans. on Circuits and Systems – II*, vol. 46, no. 6, June 1999, pp. 789–801

145. K. Lim, C.-H. Park, H. K. Ahn, J. J. Kim, B. Lim, "A Fully Integrated CMOS RF Front-End with On-Chip VCO for WCDMA Applications", *2001 IEEE Int. Solid-State Circuits Conf. Digest*, February 2001, pp. 286–287

146. "System and Building Blocks Specification and Description", Report of the IST-1999-11081 (LEMON) project, http://www.iis.ee.ethz.ch/nwp/lemon/lemon.html, June 2000

147. P. B. Kenington, *High-Linearity RF Amplifier Design*, Artech House, Norwood, 2000

148. F. H. Raab, "Intermodulation Distortion in Kahn-Technique Transmitters", *IEEE Trans. on Microwave Theory and Techniques*, vol. 44, no. 12, December 1996, pp. 2273–2278

149. "Digital Multimode Technology Redefines the Nature of RF Transmission", *Applied Microwave and Wireless*, vol. 31, no. 8, August 2001, pp. 78–85

150. K. Itho, T. Katsura, H. Nagano, T. Yamaguchi, Y. Hamade, M. Shimozawa, N. Suematsu, R. Hayashi, W. Palmer, M. Goldfarb, "2 GHz Band Even Harmonic Type Direct Conversion Receiver with ABB-IC for W-CDMA Mobile Terminal", *Proc. 2000 IEEE MTT-S Microw. Symp.*, Boston, USA, July 2000, pp. 1957–1960

151. M. Goldfarb, W. Palmer, T. Murphy, R. Clarke, B. Gilbert K. Itho, T. Katsura, R. Hayashi, H. Nagano, "Analog Baseband IC for Use in Direct Conversion W-CDMA Receivers", *Proc. 2000 IEEE RFIC Symp.*, Boston, USA, July 2000, pp. 79–82

152. M. Shimozawa, T. Katsura, N. Suematsu, K. Itho, Y. Isota, O. Ishida, "A passive-type Even Harmonic Quadrature Mixer using Simple Filter Configuration for Direct Conversion Receiver", *Proc. 2000 IEEE MTT-S Microw. Symp.*, Boston, USA, July 2000, pp. 517–520

153. A. Pärssinen, J. Jussila, J. Ryynänen, L. Sumanen, K. A. I. Halonen, "A 2-GHz Wide-Band Direct Conversion Receiver for WCDMA Applications", *IEEE J. of Solid-State Circuits*, vol. 34, no. 12, December 1999, pp. 1893–1903

154. B. Sam, "Direct Conversion Receiver for Wide-Band CDMA", *Proc. Wireless Symp.*, Spring 2000, pp. SAM-1–SAM-5

155. H. Pretl, W. Schelmbauer, B. Adler, L. Maurer, J. Fenk, R. Weigel, "A SiGe-Bipolar Down-Conversion Mixer for a UMTS Zero-IF Receiver", *Proc. IEEE Bipolar/BiCMOS Technology Meeting*, Minneapolis, USA, September 2000, pp. 40–43

156. H. Pretl, W. Schelmbauer, L. Maurer, H. Westermayr, R. Weigel, B.-U. Klepser, B. Adler, J. Fenk, "A W-CDMA Zero-IF Front-End for UMTS in a 75 GHz SiGe BiCMOS Technology", *Proc. 2001 IEEE RFIC Symp.*, Phoenix, USA, May 2001, pp. 9–12

157. W. Schelmbauer, H. Pretl, L. Maurer, R. Weigel, B. Adler, J. Fenk, "A Fully Integrated Analog Baseband IC for an UMTS Zero-IF Receiver" *Proc. AustroChip 2000*, Graz, Austria, October 2000, pp. 8–15

158. L. Maurer, W. Schelmbauer, H. Pretl, B. Adler, A. Springer, R. Weigel, "On the Design of a Continous-Time Channel Select Filter for a Zero-IF UMTS Receiver", *Proc. 2000 IEEE 51st Vehicular Technology Conference*, May 2000, Tokyo, Japan, pp. 650–654

159. J. Jussila, R. Ryynänen, K. Kivekäs, L. Sumanen, A. Pärssinen, K. Halonen, "A 22mA 3.7dB NF Direct Conversion Receiver for 3G WCDMA", *2001 IEEE Int. Solid-State Circuits Conf. Digest*, February 2001, pp. 284–285

160. W. Thomann, J. Fenk, R. Hagelauer, R. Weigel, "Fully Integrated W-CDMA IF Receiver and Transmitter Including IF Synthesizer and on-chip VCO for UMTS Mobiles", *Proc. IEEE Bipolar/BiCMOS Technology Meeting*, Minneapolis, USA, September 2000, pp. 36–39

161. W. Thomann, J. Fenk, R. Hagelauer, R. Weigel, "Fully Integrated W-CDMA IF Receiver and Transmitter including IF Synthesizer and on-chip VCO for UMTS Mobiles", *IEEE J. of Solid-State Circuits*, vol. 36, no. 9, September 2001, pp. 1407–1419

162. L. Der, B. Razavi, "A 2GHz CMOS Image-Reject Receiver with Sign–Sign LMS Calibration", *2001 IEEE Int. Solid-State Circuits Conf. Digest*, February 2001, pp. 294–295

163. M. Ugajin, J. Kodate, T. Tsukahara, "A 1V 12mW Receiver with 49dB Image Rejection in CMOS/SIMOX", *2001 IEEE Int. Solid-State Circuits Conf. Digest*, February 2001, pp. 288–289

164. S. Tabbane, *Handbook of Mobile Radio Networks*, Artech House, Norwood, 2000

165. B. McNamara, S. Zhang, M. Murphy, H.M. Banzer, H. Kapusta, E. Rohrer, T. Grave, L. Verweyen, "Dual-Band/Tri-Mode Receiver IC for N- and W-CDMA Systems Using 6"-PHEMT Technology", *Proc. 2001 IEEE RFIC Symp.*, Phoenix, USA, May 2001, pp. 13–16

166. K. Rampmeier, B. Agarwal, P. Mudge, D. Yates, T. Robinson, "A Versatile Receiver IC Supporting WCDMA, CDMA and AMPS Cellular Handset Applications", *Proc. 2001 IEEE RFIC Symp.*, Phoenix, USA, May 2001, pp. 21-24

167. J. Ryynänen, K. Kivekäs, J. Jussila, A. Pärssinen, K. A. I. Halonen, "A Dual-Band RF Front-End for WCDMA and GSM Applications", *IEEE J. of Solid-State Circuits*, vol. 36, no. 8, August 2001, pp. 1198–1204

168. J. Liu, J. Zhou, X. Zhu, J. Chen, W. Hing, "W-CDMA RF Module Design With Direct Modulation", *Proc. 2000 Asia-Pacific Microwave Conference*, December 2000, Sidney, Australia

169. T. H. Lee, S. S. Wong, "CMOS RF Integrated Circuits at 5 GHz and Beyond", *Proc. IEEE*, vol. 88, no. 10, October 2000, pp. 1560–1571

170. B. Razavi, "CMOS Technology Characterization for Analog and RF Design", *IEEE J. of Solid State Circuits*, vol. 34, no. 3, March 1999, pp. 268–276

171. J. H. Mikkelsen, T. E. Kolding, T. Larsen, "RF CMOS Circuits Target IMT-2000 Applications", *Microwaves & RF*, July 1998, pp. 99–107

172. B. Razavi, "A 2.4-GHz CMOS Receiver for IEEE 802.11 Wireless LAN's", *IEEE J. of Solid State Circuits*, vol. 34, no. 10, October 1999, pp. 1382–1385

173. B. Razavi, "A 5.2-GHz CMOS Receiver with 62-dB Image Rejection", *IEEE J. of Solid State Circuits*, vol. 36, no. 5, May 2001, pp. 810–815

174. P. G. Baltus, R. Dekker, "Optimizing RF Front Ends for Low Power", *Proc. IEEE*, vol. 88, no. 10, October 2000, pp. 1546–1559

175. http://www.iis.ee.ethz.ch/nwp/lemon/lemon.html

176. A. Springer, T. Frauscher, D. Adler, D. Pimingsdorfer, R. Weigel, "Impact of Nonlinear Amplifiers on the UMTS System", *Proc. 2000 IEEE Sixth Int. Symp. on Spread Spectrum Communications (ISSSTA)*, September 2000, Parsippany, NJ, USA, pp. 465–469

177. A. A. Abidi, G. J. Pottie, W. J. Kaiser, "Power-Conscious Design of Wireless Circuits and Systems", *Proc. IEEE*, vol. 88, no. 10, October 2000, pp. 1528–1545

178. B. A. Meyers, J. B. Willingham, P. Landy, M. A. Webster, P. Frogge, M. Fischer, "Design Considerations for Minimal-Power Wireless Spread Spectrum Circuits and Systems", *Proc. IEEE*, vol. 88, no. 10, October 2000, pp. 1598–1612

179. P. B. Kenington, "A Digital-Input, RF-Output Linearised Transmitter for 3G Base-Station Applications", *Proc. European Conference on Wireless Technology, ECWT 2001*, September 2001, London, United Kingdom, pp. 189–192

180. A. Springer, A. Gerdenitsch, R. Weigel, "Digital Predistortion-Based Power Amplifier Linearizatiopn for UMTS", *Proc. European Conference on Wireless Technology, ECWT 2001*, September 2001, London, United Kingdom, pp. 185–188

181. W. H. W. Tuttlebee, "Software-Defined Radio: Facets of a Developing Technology", *IEEE Personal Communications*, vol. 6, no. 2, April 1999, pp. 38–44

182. S. Srikanteswara, J. H. Reed, P. Athanas, R. Boyle, "A Soft Radio Architecture for Reconfigurable Platforms", *IEEE Communications Magazine*, vol. 38, no. 2, February 2000, pp. 140–147

183. J. Mitola III, *Software Radio Architecture*, Wiley, New York, 2000

184. D. Efstathiou, Z. Zvonar, "Enabling Components for Multi-Standard Software Radio Base Stations", *Wireless Personal Communications*, vol. 13, no. 1, 2000 pp. 145–166

185. W. Schacherbauer, A. Springer, T. Ostertag, C.C.W. Ruppel, R. Weigel, "A Flexible Multiband Frontend for Software Radios using High IF and Active Interference Cancellation", *Proc. 2001 IEEE MTT-S Int. Microwave Symp.*, May 2001, Phoenix, USA, pp. 1085–1088

186. W. Schacherbauer, A. Springer, C.C.W. Ruppel, R. Weigel, "A Flexible Multistandard Software Radio Frontend Concept", *Proc. European Conference on Wireless Technology, ECWT 2001*, September 2001, London, United Kingdom, pp. 265–268

Index

– downlink, 123
– factor (SF), 111, 119, 121, 149, 150, 183, 193
– uplink, 120
spurious response signal, 174, 176, 225
standard, 2, 61, 92, 94, 97–101, 146, 151, 163, 164, 181, 218, 235, 239, 240, 257, 258
standardization, 2, 91, 97, 101, 107
streaming class, 108
superframe, 111, 114
synchronization, 25, 51, 53, 60–62, 64–68, 76, 85, 105, 117, 138, 182, 187–191, 200, 203–205, 209, 210, 217
– channel, 183, 189, 192
– code, 132, 137, 138, 188, 195, 196, 198, 269, 272
– – primary, 195, 204
– – secondary, 195, 204
Synchronization Shift (SS) command, 190, 191, 204, 221

tau-dither loop (TDL), 68
TDD, 24, 25, 164, 181, 188, 215, 216
– 1.28 Mcps option, 101, 181, 189, 211, 221
– 3.84 Mcps option, 181, 211, 219
teleservices, 107, 109
time dispersion, 34, 35, 38–40, 52
time division multiple access (TDMA), 28, 29, 42, 62, 99, 186, 237
time hopping spread spectrum, 62
time switched transmit diversity (TSTD), 143, 206
timing advance, 186, 203

tracking, 65, 67, 75, 85
transmission time interval (TTI), 133, 136, 153
transmit power control (TPC), 112, 114, 140, 148, 150, 186, 190, 191, 201, 202, 206, 221
transmitter
– heterodyne, 244, 254
– homodyne, 245, 255
transport channel, 106, 109
transport format combination indicator (TFCI), 112, 114, 154, 186, 187, 190, 191, 194, 206, 209, 221

UMTS, 30, 51, 78, 89, 91, 95, 98, 101, 103, 104, 233, 237, 242, 246, 254, 256, 258
– FDD, 2, 103, 265
– TDD, 181, 271, 275
UMTS Terrestrial Radio Access Network (UTRAN), 98, 103, 106
unpaired frequency band, 210
Uplink Pilot Timeslot (UpPTS), 190, 192, 198, 203

voice activity detection (VAD), 113

W-CDMA, 2, 97, 99, 101, 103, 173, 175, 254, 256
Walsh rotator, 128, 129
Walsh sequence, 78, 79, 120
wireless communications, 1
Wireless Local Loop (WLL), 96
WLAN, 3, 61, 93, 256

Printing (Computer to Film): Saladruck Berlin
Binding: Stürtz AG, Würzburg

Return Addresses:

Interoffice Mail:
Research & Development Library
RTC 2 - 3rd floor

Courier:
Cingular Wireless
Research & Development Library
7277 164TH AVE NE
Redmond, WA 98052